Water Scarcity and Conflict in African River Basins

The book presents a critical and comparative analysis of the hydropolitical landscape of African transboundary river basins which, for much of the past century, have been affected by water scarcity.

River and lake basins can become a source of tension and conflict due to a complicated mix of environmental, demographic, diplomatic, historical, and geopolitical factors. This book; however, specifically focuses on the important, and often under looked, role played by scarcity in generating or exacerbating conflicts in shared river basins. Asserting that transboundary river basins tie states into a web of interdependence, this book raises awareness of how water scarcity, or the depletion of water resources, complicates this relationship as nations are forced to look beyond their borders to meet the demand for water to satisfy multiple needs. Taking a comparative approach, it examines three shared basins: the Orange-Senqu, Nile, and Niger River basins. While situated in different regions, all three basins are marked by serious environmental challenges that are detrimental to combustible hydropolitics over such shared water resources and they provide fascinating insights into the links between climate variability and change, water resources, human security, conflict, adaptation, and regime capacity. Overall, this book argues that conflict over transboundary resources can be prevented by the establishment of norms, rules, and the role of external actors that help regulate state behaviour and control their impacts.

This book will be of great interest to students and scholars of water resource management, hydropolitics, environmental conflict, resource scarcity, and international relations. It will also be of interest to policymakers involved in transboundary water resource governance.

Mahlakeng Khosi Mahlakeng is Lecturer in the Department of Political and Administrative Studies at the National University of Lesotho, and a Research Fellow at the University of the Free State in the Department of Political Studies and Governance. He holds a PhD in Political Science from the University of the Free State, South Africa.

Earthscan Studies in Water Resource Management
Water, Climate Change and the Boomerang Effect

Water Management in China's Power Sector
Xiawei Liao and Jim W. Hall

Reviving Indigenous Water Management Practices in Morocco
Alternative Pathways to Sustainable Development
Sandrine Simon

Transnational Corporations in Urban Water Governance
Public–Private Partnerships in Mexico and the US
Joyce Valdovinos

Defining Effective Transboundary Water Cooperation
Melissa McCracken

Corporatization and the Right to Water in Colombia
Conflicts, Citizenship and Social Inequality
Marcela López

Gender Dynamics in Transboundary Water Governance
Feminist Perspectives on Water Conflict and Cooperation
Edited by Jenniver Sehring, Rozemarijn ter Horst, and Margreet Zwarteveen

Water Management and Violent Conflict in East Africa
Scarcity and Security in Kenya and Uganda
Julia Renner-Mugono

Water Scarcity and Conflict in African River Basins
The Hydropolitical Landscape
Mahlakeng Khosi Mahlakeng

For more information about this series, please visit: www.routledge.com/books/series/ECWRM/

Water Scarcity and Conflict in African River Basins

The Hydropolitical Landscape

Mahlakeng Khosi Mahlakeng

Routledge
Taylor & Francis Group
LONDON AND NEW YORK

earthscan
from Routledge

First published 2023
by Routledge
4 Park Square, Milton Park, Abingdon, Oxon OX14 4RN

and by Routledge
605 Third Avenue, New York, NY 10158

Routledge is an imprint of the Taylor & Francis Group, an informa business

British Library Cataloguing-in-Publication Data
A catalogue record for this book is available from the British Library

ISBN: 978-1-032-43243-4 (hbk)
ISBN: 978-1-032-43245-8 (pbk)
ISBN: 978-1-003-36637-9 (ebk)

DOI: 10.4324/9781003366379

Typeset in Goudy
by Newgen Publishing UK

'From Mont-Aux-Sources in the Maluti Mountains of Lesotho to the Mediterranean Sea in Egypt…rivers run dry'

Contents

Acknowledgements

Special thanks to Professor Hussein Solomon for his continued support throughout this journey and for always believing in me. My sincerest and most humble thanks to my father, Thulo Ts'oeu Mahlakeng, for his unwavering support in seeing this book possible... Kea leboha Mofokeng! My humble thanks to my mother, Mamahlakeng 'Sis Pam' Mahlakeng, with immense love and gratitude for her support during the writing of this book. And to my partner, Nontsikelelo P. Faku, thank you for the support and time you gave, and patience you showed during the writing of this book.

For my sons: Naleli and Khanya.

1 Scarcity and conflict

By the end of the twentieth century and the beginning of the twenty-first century, the realm of international relations (IR) was characterised by the growing saliency of resource geopolitics. This pointed to the potential of conflict as a result of the scarcity of vital resources that cross political boundaries (Diehl, 1991, p.11). In recent literature, scholars have focused on the role that water plays in international affairs and international conflicts. As a result, they have supported the notion that water will be the future cause of interstate conflict (Yoffe and Wolf, 2002). Furthermore, the assumption was that the scarcity of freshwater resources will be one of the chief resource issues of the twenty-first century (Homer-Dixon, 1999a, p.15).

In 1996, the United Nations Environmental Programme (UNEP) analysis pointed out, disputes over water supplies would be the future cause of conflict among nations (Chesire, 2010, p.15). Water resources are needed for domestic and agricultural use, and to guarantee the growth of industry and investment. Consequently, in regions with a high dependence on water, competition emerges among corporations, populations, and other states. Over time, this competition over limited water resources may lead to internal and cross-border tensions and conflict (Van Schaik and Dinnissen, 2014, p. 30).

The belief that water scarcity is an obstacle to economic development and that the social stresses it causes lead to conflict are almost conventional wisdom in IR among scholars of international security (Homer-Dixon, 1999a, p. 138). Hence, water scarcity is closely linked to intra and interstate conflict (Kreamer, 2012, pp. 87–88). The United Nations (UN) identified that nine of the 13 major African river basins are susceptible to potential water-induced conflict, as being at risk of tensions. or conflict, or both (Idule-Amoko, 2009, p. 110).

For much of the past century, the hydropolitical landscape of African transboundary river basins has been affected by water scarcity (Elmi, 2002). Twumasi and Merem asserted that transboundary river basins tie states into a web of interdependence. This arrangement is further complicated by the depletion of water resources, which forces nations to look beyond their borders to meet the demand for water to satisfy multiple needs (Twumasi and Merem, 2007, p. 173). The analysis of hydropolitics refers to the politics of water, which symbolises the

DOI: 10.4324/9781003366379-1

complex relations and interactions among the states that share river basins. Such hydropolitics is a function of two variables: the rate of change in the hydrologic system, and the institutional capacity to absorb that change (Elhance, 2000, p. 202).

River water produces a potential [for] conflict between riparian states. Concerning conflict over rivers, the trend indicates that the pressures leading to a conflict are greater today than in the past due to increasing scarcity (Purkitt, 2009, p. 63). River and lake basins; therefore, can become a source of tension and conflict[1] due to the complicated environmental, demographic, diplomatic, historical, and geopolitical heat they generate (Idule-Amoko, 2009, p. 111). In basins, rivers, lakes, and water aquifers shared by more than two countries, competition for limited supplies can lead countries to see access to water as a matter of national interest. Therefore, water will increasingly become a salient element of interstate politics, in many cases even leading to violent conflict (Gleick, 1993, pp. 79–80).

Shared water resources have become a contributing factor to interstate disputes in different forms over the years. From 1950–2000 there have been 1,831 water conflicts over transboundary basins (Gleditsch *et al.*, 2006, p. 362). Shared water resources constitute a source of conflict if the river is shared across rather than along a border (Toset *et al.*, 2000, pp. 980–981). As a result, the possibility of water wars has been mooted. These are, by definition

> wars caused by the desire for access to water, in which the scarcity of water determines the means to go to war
>
> (Turton, 2000a, p. 36)

The term itself refers to

> conflict between countries, states, or groups over access to water resources, and such conflicts come as the as a result of opposing interests of water users, public or private
>
> (Tulloch, 2009)

Predictions of water wars around the globe have become a focal point of IR and have become firmly embedded in global diplomatic discourse (Du Plessis, 2000, p. 9). Despite contending beliefs regarding the occurrence of water wars, current river water shortages, the fierce competition for shared rivers, and the uneven distribution of international shared rivers constitute conditions for water conflicts (Postel and Wolf, 2001, p. 60). Similarly, according to the UN, many studies have concluded that water shortages and the competition for water are at the heart of the global security challenge and could increase the risk of conflict (UN, 2013, p. 8).

As unlikely as it seems for water to be the source of conflict; however, it bears many similarities to oil. It is essential for a wide range of human activities and it exists in relatively small amounts (Klare, 2002, p. 142). Potential water wars are already showing signs of gathering critical mass in transboundary river basins

(Simpkins, 2010). Water wars emerge within a broader framework of environmental conflicts. Environmental conflicts are,

> traditional conflicts induced by environmental degradation, and are often manifest as conflicts over resources
>
> (Libiszewski, 1992, p. 14)

These resources include water and agricultural land, and an ever-widening list of categories ranging from minerals to oil.

However, the book, intentionally, places a great emphasis and use on the notion of 'conflict' instead of 'wars'. That is, as its central focus, this book will place a great emphasis on water conflicts in contrast to water wars. For instance, the concept of war simply entails the use of force, armed conflict, fighting, or both between states or nations (Gasser, 1993, p. 2), whereas conflict encompasses the verbal, political, and violent aspects of contentions, or disagreements, or both between states or nations. Accordingly, the Australian Army Research Centre states that 'not all conflict is termed war' (Australian Army Research Centre, 2022).

The causes, implications and results of these conflicts are traditional concerns of IR (Barnett, 2000, p. 272). It is quite evident that IR, especially within broader global environmental politics, is experiencing a new landscape of global conflict for resources, with water scarcity contributing to the intensity of competition over resources (Bujra, 2002, p. 11). Water resources have historically been a source of cooperation when times are good, and conflict when resources are stretched (Abebe, 2009, p. 125).

According to Michael Klare, changes in the quantity and quality of a shared resource make disputes increasingly heated (Klare, 2002, p. 139). Countries become susceptible to conflict if the resource they are dependent on is either threatened or removed (Reyskens, 2011). However, if the capacity to absorb change is sufficient to respond to the occurring change, then the vulnerability to conflict can be minimised (Earle, 2005, p. 56).

As the demand for this fundamental and essential natural resource increases (Ashton, 2002, p. 1), water has come to be treated like oil and its protection is being insured through militaristic defence (WalesOnline, 2012). Hence, resource-based conflicts have become a major concern to international peace and security when the concept of security is expanded to include environmental security (Mathews, 1989, p. 162). This expansion of the concept of security should include the impacts of environmental stress locally, nationally, regionally, and globally (Brundtland, 1987, p. 19). Such concerns make it important to recognise that the risks are not just of a humanitarian nature, but also include political and security risks that directly affect governments (Tadesse, 2010, p. 8).

Conflicts over water resources appear to be a major source of direct international conflict. The most common environmental elements around which conflicts can erupt are population, water flow, diversion, salinisation, floods, climatic variations, and pollution (Schwartz and Sigh, 1999). Moreover, the institutional capacity to adapt to climatic variability and growing demand is inherent to the emergence of conflict.

The key to adapting to climatic variability, population growth, and the construction of large dams lies in the capacity of institutions. Low levels of institutional development have a more difficult time trying to adapt (Earle, 2005, p. 57). In addition, it is important to note that these key issues influence each other.

There is a persistent refrain in the shared waters discourse about the danger that conflicts will arise because of competition over scarce water (Muller *et al.*, 2015, p. 103). However, a substantial number of international river basins are confronted by inherent climatic variability, a natural uneven river flow, and growing demand for water resources as a result of population growth. These types of rivers characterise southern, eastern, and northern Africa. In addition, these regions have a history of political instability, a major factor that affects their institutional capacity to deal with transboundary issues. As a result, transboundary river basins in these regions and associated ecosystems could become sources of conflict (Earle, 2005, p. 56).

With an increase in ethnic and ideological tensions witnessed in interstate conflicts (Kreamer, 2012, pp. 88–89), the competition for access to vital resources has become yet another driver for conflict in IR (Klare, 2001, p. 57). The scarcity of renewable resources remains the most detrimental environmental problem (William, 1996, p. 3). The demand for scarce resources increased the likelihood of international competition over existing supplies of natural resources, which could ultimately lead to escalating tensions (Schwartz and Singh, 1999, p. 8).

This book focuses on African transboundary basins due to the continent being characterised as a continent at war with itself (Kameri-Mbote, 2005, p. 1). Moreover, the focus of the book is on water as a potential source of conflict due to water scarcity and institutional capacity, or lack thereof, or both to absorb the changes in water quality and quantity. As the book analyses the hydropolitical landscape of transboundary river basins that have been affected by water scarcity, the author acknowledges that such hydropolitics is a result of the rate of change in the hydraulic system and the institutional capacity to absorb that change.

The discourse and interest of this book in water as a potential source of conflict concentrates largely on the Orange-Senqu, the Nile, and the Niger River basins, respectively. The UN identified these basins, and six other African river basins susceptible to potential water-induced conflict, as being at risk of tensions, conflict, or both (Idule-Amoko, 2009, p. 110). The book takes a comparative perspective of these transboundary river basins and outlines foreseeable transboundary river challenges for regional security, considering the impact of environmental scarcity and regimes.

In the critical comparative analysis of these three river basins, this book uses Homer-Dixon's environmental scarcity theory, and regime theory is also known as institutional liberal theory. The purpose of this book is to determine, through Homer-Dixon's environmental scarcity theory, the impact, and effects of environmental (water) scarcity in contributing to water conflicts in these basins.

The environmental scarcity theory argues that interstate resource wars may ensue where there is a fixed or shrinking pie of natural resources. The growing population demand, climate change, degradation and depletion, and uneven distribution of water resources can act as a threat multiplier for instability in some of the most volatile regions in Africa, which are highly dependent on transboundary basins. As a result, this warrants stronger regional cooperation and effective water governance on the continent (Cooper, 2021, p. 2). In addition, UNESCO (2009, p. 2) similarly highlighted that 'the need for integrated, cooperative solutions is particularly urgent' particularly in a transboundary basin that two or more riparian states are dependent on.

In addition, the book uses regime theory to look at how regimes can mediate a nascent water conflict in these basins.

Regime theory, also known as institutional liberal theory, is used to analyse the institutional capacity in these regions to deal with, respond to, and mediate emerging environmental conflicts. Subsequently, regime theory will help in determining whether these institutional mechanisms can adapt to the changing climate and environmental landscape. It is through regime theory that we can fully comprehend why international regimes and agreements are necessary.

Regime theory assumes that once regimes, institutions, or both are in place, they can mediate between a variety of power and interest variables in a region to produce related behaviour and outcomes in an attempt to foster cooperation in 'collective security' among other things (Krasner, 1982, pp. 189–190).
To conceptualise regimes and find their relevance, a few definitions are provided. According to Keohane (1989, p. 3)

> International regimes are a major type of international institution.

Keohane and Nye (1977, p. 19) defined regimes as

> sets of governing arrangements that include networks of rules, norms, and procedures that regularise behaviour and control its effects.

North (1990, p. 3) asserted

> institutions are the formal rules of the game and are humanly devised constraints that shape interaction.

According to Hasenclever *et al.* (2000, p. 8), regimes are,

> mechanisms that help self-interested states to coordinate their behaviour.

However, a widely accepted definition of a regime is

> sets of implicit or explicit principles, norms, rules and decision-making procedures around which actors' expectations converge in a given area of IR
> (Krasner: 1982, p. 2)

According to Lindemann (2009, pp. 700–701), water regimes, on the other hand, are,

> treaties on international rivers defined as norm and rule-based cooperation for the political resolution of problems and conflict in the field of international river basin management.

The author further notes that while environmental changes may play a role in setting the stage for conflict; however, the causal relationship is highly complex, and the role of regimes is crucial in the emergence of conflict. It is not easy to predict what impact environmental scarcity will have on the levels of cooperation, conflict, vulnerability, and preservation of water resources in the region. However, the hydropolitical vulnerability to countries' environmental pressures can be reduced by developing regimes or institutions. Studying water resources and their potential to cause conflict includes not only an assessment of the rivers' natural state, territorial distribution and the changes caused due to climatic variations, and the response of regional institutions in addressing such issues.

Note

1 'Conflict' being verbal, political, or violent.

Bibliography

Abebe, L. (2009) 'Natural resource conflicts in West Africa: The case of the Niger River Basin, in Tadese, D.W. (ed.) *Climate change and transboundary water resource conflict in Africa*. Institute for Security Studies (ISS) Workshop Report. Mombasa.

Ashton, P.J. (2002) 'Avoiding conflict over Africa's resources, AMBIO', A *Journal of the Human Environment*, 31(3).

Australian Army Research Centre. (2022) What is War? Defining war, conflict and competition. Available at: https://researchcentre.army.gov.au/library/land-power-forum/what-war-defining-war-conflict-and-competition (Accessed: 5 May 2022).

Barnett, J. (2000) 'Destabilizing the environment – Conflict thesis. Review of International Studies', 26(2) April, Cambridge University Press.

Brundtland, G.H. (1987) *Our common future*. Oxford: Oxford University Press.

Bujra, A. (2002) 'African conflicts: Their causes and their political and social environment', Development Policy Management Forum (DPMF) Occasional Paper, No. 4. Addis Ababa.

Chesire, K.D. (2010) *Control over the Nile: Implications across nations*, Monterey, California: Naval Postgraduate School.

Cooper, R. (2021) *Water, climate change, and conflict*. University of Birmingham: Governance and Social Development Resource Centre (GSDRC).

Diehl, P.F. (1991) 'Geography and war: A review and assessment of the empirical literature', *International Interactions*, 17(1).

Du Plessis, A. (2000) 'Charting the course of the water discourse through the fog of international relations theory', in Solomon H. and Turton A. (eds.) *Water wars: Enduring myth or impending reality? African Dialogue*, *Monograph*, *Series 2*, KwaZulu Natal: ACCORD.

Earle, A. (2005) '*Hydropolitics in Southern Africa: What is the prognosis for peaceful development of shared watercourses?*', Forschungsinstitut Wasser und Umwelt (FWU)/ University of Siegen Research Institute for Water and Environment, Vol.3, Topics of integrated Watershed Management – Proceedings, Summer School.

Elhance, A. (2000) 'Hydropolitics: Grounds for despair, reasons for hope', *International Negotiation*, 5(2).

Elmi, A. (2002) 'Hydropolitics in the horn of Africa: Conflicts and cooperation in the Juba & Shabelle Rivers', *Proceedings of the Horn of African Conference*, Lund, 10–12 May. Available at: www.somwe.com/hydropolitics.html.

Gasser, H.P. (1993) 'International humanitarian law: An introduction', in Haug, H. (ed.) *Humanity for All: The International Red Cross and Red Crescent Movement*. Berne: Paul Haupt Publishers.

Gleditsch, N.P., Furlog, K., Hegre, H., Lacina, B., and Owen, T. (2006) 'Conflicts over shared rivers: Resource scarcity or fuzzy boundaries?' *Political Geography*, 25.

Gleick, P.H. (1993) 'Water and conflict: Fresh water resources and international security', *International Security*, 18(1).

Hasenclever, A., Mayer, P. and Rittberger, V. (2000) 'Integrating theories of international regimes', *Review of International Studies*, 26.

Homer-Dixon, T.F. (1999a) *Environment, scarcity, and violence*. New Jersey: Princeton University Press.

Idule-Amoko, A. (2009) 'The challenges of climate change and transboundary resources in Eastern Africa', in Tadese D.W. (ed.) *Climate change and transboundary water resource conflict in Africa*. Mombasa: ISS.

Kameri-Mbote, P. (2005) *Environment and Conflict linkages in the Great Lake region*. Geneva: International Environmental Law Research Centre (IELRC).

Keohane, R.O. and Nye, J.S. (1977) *Power and interdependence*. Boston: Little, Brown.

Keohane, R.O. (1989) *International institutions and state power, essays in international relations theory*. Boulder, CO: Westview Press.

Klare, M.T. (2001) 'The new geography of conflict', *Foreign Affairs*, 80(3).

Klare, M.T. (2002) *Resource wars: The new landscape of global conflict*. New York: Henry Holt and Company.

Krasner, S. (1982) 'Structural causes and regime consequences: Regimes as intervening variables', *International Organisation*, 36(2).

Kreamer, D.K. (2012) 'The past, present, and future of water conflict and international security', *Journal of Contemporary Water Research and Education*, 149(1).

Libiszewski, S. (1992) 'What is an Environmental Conflict?' *Proceedings of the first coordination meeting of Environment and Conflict Project (ENCOP)*, Zurich, 30 April– 1 May.

Lindemann, S. (2009) 'Success and failure of international river basin management: The case of Southern Africa', in Brauch, H.G. et al. (eds.) *Facing global environmental change: Environmental, human, energy, food, health and water security concepts*. Hexagon Series on Human and Environmental Security and Peace, Vol.4. Springer - Verlag Berlin Heidelberg.

Muller, M., Chikozho, C. and Hollingworth, B. (2015) 'Water and regional integration: The role of water as a driver of regional economic integration in Southern Africa', Report to the Water Research Commission (WRC), Report No 2252/1/14, March.

North, D. (1990) *Institutions, institutional change and economic performance*. Cambridge: Cambridge University Press.

Postel, S, and Wolf, A.T. (2001) 'Dehydrating Conflict'. *Foreign Policy*, 126(Sept–Oct).

Purkitt, H.E. (2009) *African environmental and human security in the 21st century*. Amherst, New York: Cambria Press.

Reyskens, M. (2011) 'In hot pursuit: Resource wars in Africa', 16 November. Available at: www.consultancyafrica.com/index.php?option=com_content&view=article&id=901:in-hot-pursuit-resource-wars-in-africa-&catid=60:conflict-terrorism-discussion-papers&Itemid=265. (Accessed 19 April 2013).

Schwartz, D. and Singh, A. (1999) *Environmental conditions, resources, and conflicts: An introductory overview and data collection*. Nairobi: The United Nations Environment Programme (UNEP).

Simpkins, G. (2010) 'Africa faces water war threats', Africa Rising 21st Century, BlogSpot, 26 February. Available at: http://africarising2010.blogspot.co.za/2010/02/africa-faces-water-war-threats.html. (Accessed: 11 March 2016).

Tadesse, D. (2010) The impact of climate change in Africa, Institute for Security Studies, Paper 220, November.

Toset, W., Hans P., Gleditsch, N.P., and Hegre, H. (2000) 'Shared rivers and interstate conflict,' *Political Geography*, 19.

Tulloch, J. (2009) 'Water Conflicts: Fight or Flight?' Allianz Knowledge, August 26. Available at: http://knowledge.allianz.com/en/globalissues/climate_change/natural_disasters/water_conflicts.html (Accessed: 11 May 2013).

Turton, A.R. (2000a) 'Water wars in Southern Africa: Challenging conventional wisdom', in Solomon, H. and Turton, A. (eds.) *Water wars: Enduring myth or impending reality?* African Dialogue, Monograph, Series 2, KwaZulu Natal. Durban: The African Centre for the Constructive Resolution of Disputes.

Twumasi, Y.A. and Merem, E.C. (2007) 'Using remote sensing and GIS in the analysis of ecosystem decline along the River Niger Basin: the case of Mali and Niger', *International Journal of Environmental Research and Public Health*, 4(2).

UN. (2013) 'Water Security and the Global Water Agenda: A UN-Water Analytical Brief', Ontario: Institute for Water, Environment and Health (UNU-INWEH).

UNESCO. (2009) From potential conflict to co-operation potential: Water for peace. Available at: http://unesdoc.unesco.org/images/0012/001262/126261e.pdf. (Accessed: 2 August 2017).

Van Schaik, L. and Dinnissen, R. (2014) 'Terra incognita: Land degradation as underestimated threat amplifier', Clingendael Report.

Wassenaar: Netherlands Institute of International Relations.

WalesOnline. (2012). 'Treat water like oil and sell it for profit, says ex-DwrCymru Chief', 10 April. Available at: www.walesonline.co.uk/news/wales-news/treat-water-like-oil-sell-2047334. (Accessed: 30 July 2013).

William, E.D. (1996). *Environmental scarcity as a cause of violent conflict*. Kansas: School of Advanced Military Studies, United States Army Command and General Staff College.

Yoffe, S. B and Wolf, A. T. (2002) *Water, conflict and co-operation: Geopolitical perspectives*. Available at: www.transboundarywaters.orst.edu/publication/geographical_perspectives/. (Accessed: 4 September 2013).

2 Homer-Dixon's environmental scarcity theory and regime theory

Since the 1990s, much conceptual and theoretical debate has focused on the assumed violent outcome of environmental stress (Gleditsch, 1997; Diehl and Gleditsch, 2001). The Toronto[1] and Zürich-Bern[2] groups produced many empirical insights on the linkages between environmental scarcity, environmental stress, and environmental conflict (Tanzler *et al.*, 2002, p. 61).

The link between the environment and conflict has been a subject of discussion for a while (Kameri-Mbote, 2005, p. 2). Various research aspects of the environmental conflict have been critically received and discussed. A number of unresolved dilemmas emerge from these debates that characterise the literature on environmentally-induced conflict (Hagmann, 2005, p. 11).

In policy circles, scarcity-related phenomena are often related to political unrest and conflicts. Environmental stresses, therefore, tend to receive more attention on the political agenda, especially to create a sense of urgency to address environmental issues and scarcity phenomena. Some critics say this is unjustified and they point out that evidence of the relationship between environmental stresses and insecurity is far-fetched. Instead, they refer to other conflict risks, such as economic, socio-political and demographic factors (Van Schaik and Dinnissen, 2014, p. 12).

Hendrix and Salehyan (2010, p. 15) proposed that it was the interaction between environmental stress with social and political factors that were key to whether water-related conflict developed or not. Similarly, Raleigh and Kniveton (2012, p. 52) emphasised that conflict is complex, sometimes linked to resources, but often fundamentally rooted in other factors such as climatic, socioeconomic, political, and institutional factors. According to Kameri-Mbote (2005, p. 2), Homer-Dixon (1991; 1994; 1998; 1999a; 1999b) is the foremost academic advocate of environment and conflict linkages. Homer-Dixon developed models to demonstrate the linkages between renewable resources scarcity and conflict. In addition, these models have generated deep concern for environmentally-induced conflict (Homer-Dixon and Blitt, 1998, p. 1).

Furthermore, Homer-Dixon (1994; 1999a; 1999b) argued strongly for the existence of a relationship between conflict threats and environmental factors, notably scarcity of natural resources. In many parts of the developing world, environmental scarcity, especially land and water, is worsening rapidly. This

DOI: 10.4324/9781003366379-2

worsening scarcity has immediate direct, indirect, or both impacts on the well-being of people dependent on these resources (Homer-Dixon, 1999a, p. 53).

Homer-Dixon environmental scarcity theory and regime theory

Supporting the environment-conflict argument is Homer-Dixon's environmental scarcity theory, which argues that the scarcity of renewable resources such as cropland, fish, forests and, most importantly, water could lead to violent conflicts (Homer-Dixon, 1999a, p. 8). The environmental scarcity theory illustrates how the scarcity of renewable resources can contribute to social breakdown and violence (Homer-Dixon, 1999a, p. 4).

Environmental scarcity refers to the declining availability of renewable natural resources such as freshwater or soil (Bingham, 2001).

This means that the supply of renewable resources such as water is not sufficient to meet the local demand (UNEP, 2012, p. 29). Environmental scarcities usually have complex causes. The depletion and degradation of a resource are a function of the physical vulnerability of the resource, the size of the population consuming the resource, the consumption behaviour, and the distribution of the resource.

The environmental scarcity theory stipulates that population growth, the depletion and degradation of a resource, and the uneven distribution of a resource are not mutually exclusive; they often occur simultaneously and interact (Homer-Dixon, 1994, pp. 8–11). Uneven distribution never acts on its own, its impact is always a function of its interaction with rapid population growth and resource degradation and depletion (Schwartz *et al.*, 2000, p. 80).

Homer-Dixon's environmental scarcity theory found that problems of declining resource supply and rising resource demand were always intimately entangled with uneven resource distribution and were environmental factors capable of triggering environmental conflict (Schwartz *et al.*, 2000, p. 80). However, the theory clearly states that population pressure, environmental degradation, unequal access to a resource alone, or both cannot lead to violent outcomes.

With the interaction of these causes of scarcity come four principal social effects that have far-reaching consequences on states and people. These include decreased agricultural potential; regional economic decline; population displacement; and the disruption of legitimised and authoritative institutions and social relations (Homer-Dixon, 1991, p. 91).

First, environmental change (in the form of supply-induced scarcity, the depletion and degradation of environmental resources such as water, or both) reduces or constrains agricultural productivity. Worsening deforestation and scarcity of water have huge effects on agricultural production; therefore, affecting food outputs. For example, deforestation changes hydrological cycles by affecting transpiration rates, soil moisture, and precipitation patterns. This will result in erosion and silting, and ultimately lead to incidences of drought and flooding that constrain irrigation capacity and regional productivity. Deforestation also results in degraded and eroded soils, which affects root depth. This makes plants vulnerable to drought and eventually affects agricultural productivity.

However, the overuse and pollution of water supplies can also result in constrained irrigation capacity and regional productivity (Homer-Dixon, 1999a, p. 81). Homer-Dixon (1999a, p. 87) further suggests that institutions are largely helpful in assisting countries and their agricultural systems to respond to, adapt, or both effectively to rising land and water scarcities. If these institutions fail, then the availability of water will decrease, and agricultural systems will remain inefficient.

Second, regional economic declines can be caused by supply-induced scarcities such as environmental degradation and depletion. This form of environmental scarcity may negatively affect the economy of a region, either directly or indirectly through other social effects such as changes in agricultural productivity. Consequently, environmental scarcity can depress the economy's long-term productivity, which may have a large effect on a country's overall economic development (Homer-Dixon, 1999a, p. 89).

Deforestation leads to water shortages because of erosion and silting. This may disrupt hydropower production and river volumes; therefore, simultaneously reducing or constraining regional agricultural and economic productivity. The resulting overuse and pollution of water supplies can also affect and constrain economic productivity by giving rise to waterborne human and animal diseases such as cholera. This is an indication that a variety of environmental stresses could affect wealth production (Homer-Dixon, 1999a, pp. 88–89).

Third, Homer-Dixon (1999a, p. 93) argued that environmental scarcity can also lead to massive population displacements, migrations, or both to neighbouring regions. This is evident in the effects of the three sources of scarcity. Demand-induced scarcity, brought about by population growth, is a direct and major cause of population displacement as population growth critically reduces per capita access to water in a specific area; therefore, forcing people to move (Homer-Dixon, 1999a, p. 95).

Supply-induced scarcity, brought about by water degradation and depletion, also causes population movements. The reduced water output, combined with population growth, places immense pressure on already limited per capita access to water, forcing people to move. Structural-induced scarcity, brought about by the unequal distribution of water, produces a wide gap between areas with and without water, eventually creating gaps between people's current satisfaction and potential satisfaction elsewhere. As a result, this may also lead to people moving to these perceived areas of satisfaction (Homer-Dixon, 1999a, pp. 95–96).

Lastly, the failure of authoritative institutions to help societies adapt to scarcities also raises the probability of violence. This is because weak state institutions reduce the ability of societies to generate and deliver the social and technical ingenuity needed to respond to environmental scarcity (Homer-Dixon, 1999a, p. 98). The failure of these institutions is evident in their inability to improve, build, or both infrastructure such as wells, dams, canals, pipelines, irrigation systems, and large facilities to control industrial and municipal pollution. Institutions are required to respond to the increasing financial and political demands on the state to deal with severe water scarcity (Homer-Dixon, 1999a, p. 101).

Severe environmental scarcity can also exacerbate divisions among groups sharing a particular resource. Scarcity increases the gap between winners and losers, for example, between groups that gain from scarcity and those that suffer from it. This gap encourages competition among groups for the control of resources critical for survival. In addition, the divisions and gaps among groups are the social effects of environmental scarcity that can also aggravate social relations, which can disrupt institutions (Homer-Dixon, 1999a, p. 96). These four social effects, either individually or in combination, produce or exacerbate conflict (Homer-Dixon, 1994, p. 6).

Moreover, the three sources of scarcity have a special political significance and can interact and reinforce each other in extraordinary ways. If these three sources of scarcity interact, conflict is inevitable. The first aspect of this interaction, *resource capture*, is very significant in the case of a shared resource. Resource capture takes place when powerful groups or states recognise that a key resource is becoming more scarce (due to both supply and demand pressures) and use their power to shift the laws and institutions governing resource access in their favour. This shift ultimately imposes severe structural scarcities on weaker groups (Homer-Dixon, 1999a, p. 15). Another motivation for resource capture is greed. Securing water resources at the expense of other riparian countries simply means that this will open opportunities (Homer Dixon, 1999a, p. 74).

The second aspect of this interaction is *ecological marginalisation* (Homer-Dixon, 1999a, p. 15). Environmental scarcity will produce large numbers of environmental refugees[3] (Homer-Dixon, 1999a, p. 93). Unequal resource access can combine with population growth to cause migrations to regions that are ecologically fragile such as steep upland slopes, areas at risk of desertification, and tropical rain forests. High population densities in these areas, combined with a lack of knowledge and capital to protect local resources, cause severe environmental damage, chronic poverty and violent disputes (Homer-Dixon, 1999a, p.15).

In explaining how the scarcity of renewable resources can contribute to social breakdown and violence, the book will provide environment-conflict linkages.

Homer-Dixon (1999a, p. 4) claims that the idea and work laid by scholars[4] such as Norman Myers (1996; 2004) to explain the link between the environment and conflict was speculative and imprecise, and was marked by an absence of empirical and theoretical structure.

The assumption behind Homer-Dixon's environmental scarcity approach is,

> resource scarcity, through the three causal forms of scarcity (i.e., population growth, degradation and depletion of resources, and uneven distribution of resources), have the potential to cause conflict
>
> (Homer-Dixon, 1999a, p. 177)

The arguments of the book are central to this assumption. The concept of environmental scarcity encompasses all three of these sources. Unfortunately, most analysts often study resource depletion and population growth in isolation from resource distribution.

The environmental scarcity theory allows for the incorporation of these three distinct sources of scarcity into one analysis. The incorporation of all these three distinct sources of scarcity into one analysis is helpful in the research of transboundary river basin disputes. Hence, the theory argues that an analysis of transboundary river basin disputes should not be made only based on uneven distribution, but concurrently with water demand and supply (Homer-Dixon, 1994, pp. 8–11). Empirical evidence suggests that the first two sources are most dangerous when they interact with unequal resource distribution (Homer-Dixon, 1994, p. 8; Urdal, 2008, p. 593). Scarcity, driven by the ongoing processes of environmental degradation and escalating population growth, is believed to be rapidly increasing in many marginal environments.

There are concerns that this combination of resource depletion and increasing demand may increase competition and conflicts (Munia *et al.*, 2016, p. 1). In many regions, resource availability is being squeezed by supply and demand pressures. The scarcity of water as a vital natural resource guarantees future social unrest, geopolitical friction, and war (Klare, 2013). According to Ohlsson (1999, p. 211), the driving force for conflict over water between countries is the desire to increase supply and manage demand. Environmental effects may cause social effects that, in turn, could lead to conflict (Homer-Dixon, 1991, p. 79).

Population growth affecting water resources and a nascent conflict

Demand-induced scarcity is primarily caused by population growth. If a resource base is constant, the availability of resources per person will diminish with the increasing number of people that have to share it. Such scarcity can also arise from an increase in demand per capita (Urdal, 2008, pp. 592–593). According to Homer-Dixon (1999a, p. 51), demand-induced scarcity is a result of population size multiplied by per capita demand for a given resource, and an increase in either population or per capita demand increases total resource demand. For instance, the number of people living in an arid region might increase, which, all other things being equal, will increase the total demand for resources such as water for irrigation.

Homer-Dixon (1999a, p. 52) argues that the growing population and greater per capita resource demand can simultaneously boost demand and supply-induced scarcity. For example, increased population size and per capita demand for a given resource can have dual effects on environmental scarcity. On the one hand, they both increase the demand for the resource, and they can both decrease supply by contributing to resource depletion and degradation (Homer-Dixon, 1999a, p. 52).

Homer-Dixon (1999a, p. 48) indicated that an increase in demand assumes that the growth in population divides the pie into smaller slices for each individual, group, or state. Demand-induced scarcities arise only with resources that are rivalrous (e.g., fisheries, cropland, forests, and water). Further, a resource is deemed to be rivalrous when its use by one economic actor reduces its availability for others (Homer-Dixon, 1999a, p. 8; Percival and Homer-Dixon, 1998, p. 280). A demand-induced scarcity results from the water needs of an increasing population, which justifies the demand for increased welfare (van der Molen and Hildering, 2005, pp. 134–145).

The demand-induced scarcity argument posits population growth or an increase in consumption levels that decrease the amount of limited natural resources available to each individual (Bingham, 2001). Regionally and globally, population growth and rising resource consumption affect resource availability (Homer-Dixon, 1999a, p. 112). Homer-Dixon (1999b, p. 61) argued that as the population increases, so will the scarcity of renewable resources. In many cases, population growth does not damage the environment, but often this growth, in combination with prevailing social structures and consumption levels makes environmental degradation and depletion worse (Homer-Dixon, 1999a, p. 55).

The biggest of these challenges relates to water for production. This includes power generation, and industrial and agricultural production to support the rapidly increasing populations in riparian states that must be sustained through adequate food security (Idule-Amoko, 2009, p. 111).

Human activities have a huge impact on renewable resources. Physical trends that are evident and detrimental to renewable resources include human population, rising energy consumption, global warming, ozone depletion, deforestation, and rising scarcity of fresh water (Homer-Dixon, 1999a, pp. 52–53). Environmental threats have often been attributed to such rapid human population growth (UNEP, 2006, p. 1). The continuation of the downward spiral of increasing population and declining environmental quality is inevitable (Urdal, 2005, p. 418). Water has been regarded as finite and fixed because of the rapid population growth. The dangers that come with international competition for adequate water resources will inevitably increase (Klare, 2001, p. 57).

Due to the lack of sufficient water for growing populations, Africa will be confronted with greater disputes and tensions in the years ahead (Simpkins, 2010). As argued by the resource scarcity perspective, population growth and density may lead to a scarcity of renewable natural resources such as productive land, fresh water, and forests (Weiner and Russell, 2001, p. 3). Roudi-Fahimi *et al.* (2002, p. 4) asserted that population growth was the most important trend that affects the water demand because it directly increases the pressure over available resources and production capabilities in a country. This is due to water being needed for more than just the basic natural use of human consumption (Roudi-Fahimi *et al.*, 2002, p. 4). However, the real challenge lies in the availability of water for large-scale irrigation and food production to feed the people or to obtain revenue from agricultural exports. Therefore, population growth in a country not only increases the demand for domestic water consumption but also for large quantities of food, which requires considerable amounts of water (Hernandez, 2012, pp. 2–3).

In recent years, the global pressure on limited freshwater resources has been mounting, driven by an increase in population growth. In October 2011, the global human population surpassed 7 billion and is projected to rise to 8 billion by 2025. This increase, coupled with rising rates of consumption, places further demands on the supply of renewable resources. As demand increases, some countries are already reaching the limits of their water resources. This has been made worse by phenomena such as El Nino[5], which has acutely affected sub-Saharan Africa with a decrease in rainfall. Rainfall and river flow in Africa display high levels of variability across a range of spatial and temporal scales, with important consequences for the management of water resource systems (Conway *et al.*, 2009, p. 42). As a result, competition for water intensifies between countries. This makes water an increasingly politicised issue (UNEP, 2012, pp. 17–18).

Weiner and Russell (2001, p. 3) argued that resource scarcity is assumed to lead to increased intergroup competition, and, under unfavourable economic and political conditions, such competition can take the form of violent conflict. Consistently, Goulden and Few (2011, p. 25) found that population growth and density were associated with this increased risk of conflict. According to Kennedy (2001), population growth engenders resource scarcity by creating a demand-induced scarcity, which forces states to adopt greedy measures in an equation where a resource is shared among more than one state. A comprehensive argument is that population growth will outstrip the natural resources of the immediate environment, leading to deprivation and, ultimately, to conflict and instability either directly through competition for scarce resources, or indirectly by generating environmental refugees (Barnett, 2000, p. 278).

In shared river basins, when the water demand outstrips the available supply, a nation can justify military action (whether offensive or defensive) in the name of economic preservation and national security. Population growth may also lay the foundation for conflict in the absence of effective regimes (Van Schaik and Dinnissen, 2014, p. 33). In this book, effective regimes[6] refer to properly functional governance systems such as institutions, agreements, treaties or both meant to contribute to solving international problems. In the instance of environmental problems, this would require functional environmental agreements (Wettestad, 1999, p. 173; Young, 2011, p. 19854). Rapid population growth and density are often cited as significant conflict risk factors. The risk for interstate conflict increases when the reinforcing nature of population growth and environmental stresses coincides with the weakening of regimes (Van Schaik and Dinnissen, 2014, p. 33). The reason behind such an assertion is that the scarcity of renewable resources alone is neither a necessary nor sufficient cause of conflicts. Rather a number of contextual factors such as political, economic, and social factors play a major role in inciting these types of conflict between states.

Degradation and depletion of water resources and a nascent conflict

Central to Homer-Dixon and Blitt's (1998, p. 6) analysis of the environment and conflict linkages is the idea that environmental change is only one of three primary sources of renewable resource scarcity. When environmental conflicts break out, it is usually against the background of a number of different factors interacting with one another. A number of scholars have asserted that large-scale human-induced environmental pressures may seriously affect national and international security (Homer-Dixon, 1991, p. 76).

The total effect of human activity on the environment in a particular ecological region is mainly a function of two variables: first, the product of the total population in the region and physical activity per capita; and second, the vulnerability of the ecosystem in that region to those particular activities. Activity per capita, in turn, is a function of available physical resources (which include renewable resources such as water, forests, and agricultural land) and ideational factors, including institutions (Homer-Dixon, 1991, p. 79).

Kameri-Mbote (2005, p. 2) presupposed that the degradation or depletion of land and water will make a greater contribution to turmoil in the coming decades than climate change and ozone depletion. Degradation and depletion of resources can increase the risk of conflict in areas where countries that are highly dependent on natural resources are experiencing high rates of deforestation and soil degradation and in those with low per capita availability of arable land and fresh water (Homer-Dixon, 1999a, p. 53).

We can narrow the scope of this research problem by focusing on how environmental change affects conflict. It may have different causal roles: in some cases, it may be a proximate and powerful cause; in others, it may only be a minor and distant player that involves many political, economic and physical factors (Homer-Dixon, 1991, p. 76).

Supply-induced scarcity results from the degradation or depletion of natural resources (Urdal, 2008, p. 593). According to supply-induced scarcity, a drop in the supply of a key resource assumes that the resource pie shrinks because there has been a reduction in quantity and quality (Homer-Dixon, 1999a, p. 48). This scarcity refers to environmental degradation that decreases the overall available amount of a limited natural resource; therefore, decreasing the amount available to each individual (Bingham, 2001).

Environmental degradation, the chapter relates much of its arguments to the negative impact on nature, human beings, and society. This refers to the human-made environmental changes that consequently negatively impact human society (Libiszewski, 1992, p. 4). On the other hand, Homer-Dixon argued that natural resources may also be degraded and depleted from causes that are not human-induced, such as natural disasters or less dramatic natural variation. Of the major environmental changes facing humankind, degradation and depletion of agricultural land, forests, water, and fish will contribute more to social turmoil in the coming decades than climate change or ozone depletion (Urdal, 2008, p. 593).

Resource depletion and degradation are only one of the three sources of environmental scarcity, as seen earlier. Depletion and degradation produce a decrease in the total resource supply (Homer-Dixon, 1999a, p. 15).

Supply-induced scarcity results from rivers running dry, lowered water tables, and polluted groundwater and surface water courses (van der Molen and Hildering, 2005, p. 135).

It simply becomes less of a resource because of nonsustainable use that does not allow the resource to regenerate. For example, pastoralism, agriculture, and irrigated farming ultimately lead to the erosion of the landscape (Baechler, 1998, p. 69).

Environmental security literature stipulates that supply-induced scarcity in the form of environmental degradation is a key driver of civil violence around the world (Homer-Dixon, 1999a, pp. 63–77). The main idea, proposed by Homer-Dixon, is that through environmental degradation resources become scarce and people start fighting over them. According to Benjaminsen (2008, pp. 819–21), African dry lands are among the areas most seriously affected by degradation.

Countries may fight over dwindling supplies of water and the effects of upstream pollution (Gleick, 1989, p. 336). Water supplies that underpin the livelihoods of people are renewable. Renewable resources are replenished over time by natural processes. In most cases, if used prudently, they should sustain an adequate standard of living indefinitely. Unfortunately, in many regions where people rely on renewable resources, they are being depleted or degraded faster than they are being renewed (Homer-Dixon, 1999a, p. 13).

Unlike many analyses, the environmental scarcity theory interprets all types of environmental depletion or damage as various forms of the scarcity of renewable resources. It views land degradation, climate change and migration as fundamental causes of the increase in the scarcity of land and water resources, and as sources that increase the scarcity of the regular patterns of rainfall and temperature on which communities and states rely (Homer-Dixon, 1999a, p. 9).

Land degradation

Land degradation can be defined as

> a long-term loss of ecosystem function and productivity caused by disturbances from which the land cannot recover
>
> (George and Cowie, 2011, p. 370)

In the literature, the term land degradation is mostly used to describe the process of land losing its fertility, and to describe situations where land has lost all of its fertility and has, for instance, turned into a desert (Abdi *et al.*, 2013, p. 42).

According to Van Schaik and Dinnissen (2014, p. 12), the availability of water, a key resource for survival, can diminish with land degradation and this can pose serious security implications. Although water scarcity and its link to land degradation are increasingly on the political agenda, Van Schaik and Dinnissen (2014, pp. 11–31) assumed that land degradation was an underestimated threat and environmental challenge.

They further asserted that due to a lack of knowledge and openness, it was difficult to determine the precise magnitude and influence of land degradation on food, water scarcity, and conflict risk. This also undermined the sense of urgency to implement concrete measures to counter land degradation. These measures are necessary to tackle negative effects and to stimulate the restoration of degraded lands. Given the widespread and often negative consequences of land degradation, such as migration risks and their implications, several multilateral, national, and local initiatives have tried to raise land degradation as an issue of concern (Van Schaik and Dinnissen, 2014, p. 12).

Land degradation is caused by a combination of human activity and nonhuman factors. Human activities include deforestation and pollution of land and water. These factors are often referred to as bad land management practices and are sometimes related to weak governance, war, or other forms of conflict. Nonhuman factors include rainfall variability, wind, and temperature changes. These changes in weather conditions can be related to climate change.

Land degradation, through unsustainable agriculture, industry, and domestic use are paramount to the depletion and degradation of transboundary river basins. Since land degradation reduces people's ability to use the land for agricultural production, it potentially reinforces scarcity issues. Land degradation is caused by the long-term unsustainable use of agricultural and forestry land, the exploitation and poor management of land, or both. In this way, ecosystems lose their functions and the quality and productivity of agricultural land decrease. The quality of the land may also decline because of climate change or local environmental pollution. Subsequently, this may negatively impact food supplies and the ability of the land to retain water (Van Schaik and Dinnissen, 2014, pp. 11–12).

Land degradation is inextricably linked to the quality, volume, and timing of water flow in basins. Land degradation can have negative consequences for the availability, quantity, and quality of water resources. It decreases the ability of the land to capture water at levels accessible to humans (i.e., not too deep below the surface) and can contribute to water pollution (i.e, if land degradation is caused by pollution). Hence, when land becomes degraded, the availability of (renewable) water is likely to decline, leading to reduced access to drinking water (Baecher *et al.*, 2000, p. 73).

Land degradation has become a huge challenge within and between riparian countries (Di Nunzio, 2013, p. 6). Over the years, a considerable number of studies have been published analysing whether, how, and the extent to which environmental degradation may cause violent intra and interstate conflicts at local, subnational, and national levels. Under certain circumstances, land degradation may aggravate tensions in society that could lead to violent conflict (Van Schaik and Dinnissen, 2014, p. 31).

Notwithstanding healthy criticism regarding alarmist statements, the relationship between land degradation and conflict does exist. However, this relationship is a complex one and is indirect (Van Schaik and Dinnissen, 2014, pp. 11–12). Land degradation is more directly related to conflict risk factors.

Climate change

The sharing of international rivers in Africa in the face of climate change might lead to regional security tensions. Climate change simply refers to a change in global or regional climate patterns. Such irregular weather patterns, influenced by heavy rainfall in some regions and strong sunshine in others, lead to environmental degradation by lowering the environmental conditions necessary for human survival (Tsuma, 2011, p. 5). Climate change introduces uncertainty in the management and preservation of water resources in that it shifts precipitation patterns and the overall water supply (Homer-Dixon, 1999a, p. 67).

Climate change is also one of the many security threats facing Africa because it is likely to produce a host of social and political problems that could weaken states and societies. Climate change may thus seriously threaten political and economic stability in Africa. It may also put a severe strain on the capacities of states and societies to coordinate activities, communicate and organise (Tadesse, 2010, p. 7). According to Swain and Krampe (2011, p. 17).

> as per the Intergovernmental Panel on Climate Change (IPCC) findings, two regional patterns are likely for Africa. Northern and southern Africa will become much hotter (minimum plus 4°C) and drier (about 10–20% less rain).

Climate change has moved up the political agenda over the last two decades. This is because it has socioeconomic, political, and demographic (through migration) impacts, which many fear will lead to increased tensions and violent conflict (Leroy and Gebresenbet, 2011, p. 9). Since the 1980s, a growing body of research has dealt with the relationship between environmental stress (e.g., environmental degradation and resource scarcity) and conflict. More specifically, evidence is mounting that the adverse effects of climate change can, particularly by interacting with a number of other socioeconomic factors, contribute to an increased potential for conflict (Tanzler *et al.*, 2002, p. 4).

During the same period, the issue of climate change increased in importance on the agenda of international environmental politics. This witnessed numerous international processes on climate change being established, such as the IPCC in 1988, the Second World Climate Conference in 1990, and the adoption of the UN Framework Convention on Climate Change in 1992. In recent years, climate change has become a major topic in international affairs due to the negotiations of the Kyoto Protocol, which was finalised through the Marrakesh Accords (Tanzler *et al.*, 2002, p. 116).

According to Leroy and Gebresenbet (2011, p. 10), the fear and concern of climate-induced conflicts became evident in 2007, when the African Union and the UN held their first debates on the security implications of climate change. A substantial body of qualitative research tried to establish a link between climate change and conflict. For instance, physical manifestations of climate change may lead to various socioeconomic and political impacts and, if not well managed, may eventually lead to violent conflict (Leroy and Gebresenbet, 2011, p. 11).

Bronkhorst (2011, p. 31) found that there is already broad consensus that the environment and climate can contribute to conflict, or be an underlying trigger (Bronkhorst, 2011, p. 31). Climate change, one of the factors causing or aggravating land degradation, is often referred to as a *threat multiplier* because it exacerbates existing trends, tensions, and instabilities. Given the history of resource and political conflicts in Africa, climate change could aggravate territorial and border disputes and complicate conflict resolution and mediation processes. Constraints on water availability are a growing concern in Africa, and climate change will exacerbate this situation (Tadesse, 2010, pp. 7–8).

The German Environment Ministry (2002, p. 4) also found,

> the adverse effects of climate change can, particularly by interaction with a number of socio-economic factors, contribute to an increasing potential for conflict.

Similarly, Homer-Dixon (1999a, p. 14) asserted that climate change is most likely to have major effects on societies, not by acting as an isolated environmental pressure, but by interacting with other long-present resource pressures, such as stressed water supplies.

Climate change and its consequences, such as floods, droughts, and an increase in extreme weather events, may not directly cause conflict but may trigger or aggravate it. The results of droughts and floods will pose regional security challenges in Africa (Swain and Krampe, 2011, p. 17). The core challenge is that climate change threatens to overburden states and regions that are already fragile and conflict-prone. Countries predicted to be the worst affected by climate change, are also affected or threatened by violence and instability. In these countries, the consequences of climate change are most likely to combine with other possible factors contributing to violent conflict, for example, poverty, bad governance, and the legacy of past conflicts all put additional strain on fragile social and political systems (Tadese, 2010, p. 8).

Water is a key adaptation challenge to climate change (Ahmed, 2011, p. 26). During other research that focused on the relationship between climate change and security, scholars looked at various effects of climate change. Some scholars analysed rain precipitation patterns, while others looked at the effects of climate change on water scarcity and conflict. Climate plays an important role in the evolution of river channel systems (Grainger and Conway, 2014, p. 835). The impact of climate change will have a clear bearing on access to shared water resources as it affects hydrological cycles from global to local levels (Swain and Krampe, 2011, p. 17).

Declining water resources and diminishing arable land are already intensifying competition for these resources and creating tensions among displaced populations (Tadese, 2010, p. 8). A third dimension for investigating the climate change–conflict nexus is that of migration. This is because the threats brought by climate change and environmental disasters can induce forced migration and produce competition among communities and nations for water and basic needs resources. For instance, conflict may arise over water resources when forced migration comes in contact with communities and nations that are already struggling to access scarce water resources (Tadesse, 2010, p. 7).

Migration

Migration can be expected to either aggravate or mitigate the relationship between land degradation and scarcity; therefore, leading to conflict. Migration is often quoted as a source of conflict, as it increases competition over resources, which may be scarce. The risk of migration increasing tensions is highest when poor people migrate to already poor and unstable countries (Williams and Gemenne, 2016, p. 154).

In 1999, the International Red Cross reported that 25 million people had been displaced by environmental disasters. In 2009, the UN High Commissioner on Refugees (UNHCR) estimated that number was 36 million, 20 million of whom were listed as victims of climate change-related issues (Lam, 2012a). According to Lam (2012b), the number of environmental refugees is expected to grow to 50 million by 2020. It has been estimated that the figure will triple to 150 million by 2050. The UNHCR also believes that the population of the environmentally displaced has already far outstripped the number of political refugees worldwide, which is currently at around 10.2 million.

According to the UN *International Migration Report* (2015, p. 1), the number of international migrants worldwide has continued to grow rapidly over the past 15 years reaching 244 million in 2015, up from 222 million in 2010 and 173 million in 2000. Nearly two-thirds of all international migrants live in Europe (76 million) or Asia (75 million). Northern America hosted the third largest number of international migrants (54 million), followed by Africa (21 million), Latin America and the Caribbean (9 million), and Oceania (8 million).

Migration is motivated by various factors such as economic decline, poverty, natural disasters, floods, land degradation, social unrest, or war (Werz and Conley, 2012, p. 4). Water resources have also been deliberately used to foment migration. In Nigeria and Sudan, for instance, water resources such as village wells of one group have been bombed, poisoned, or polluted by another group dumping bodies in them as a vehicle to influence migration (Kreamer, 2012, pp. 89–90).

The link between migration and conflict can be exacerbated by the sudden resource scarcity caused by climate change (i.e., extreme weather events) (Werz and Conley, 2012, p. 4). The effects of climate change will likely contribute to an increase in forced migration. Climate-related migration, due to reduced water availability, may increase the risk of political instability and conflict between different users (Nordas and Gleditsch, 2007, p. 630). Climate migrants are,

> those people who have been forced to leave their traditional habitat, temporarily or permanently, because of a marked environmental disruption[7] (natural and/or triggered by people) that jeopardized their existence and/or seriously affected the quality of their life
>
> (El-Hinnawi, 1984, p. 4)

Uneven distribution of water resources and a nascent conflict

The third kind of scarcity is uneven resource distribution, which is sometimes referred to as structural-induced scarcities (Homer-Dixon, 1999a, p. 52). In resource capture, demand- and supply-induced scarcities interact to produce structural scarcities (Kameri-Mbote, 2005, p. 2). This scarcity is caused by

> unequal distribution that concentrates a resource in the hands of some groups and subjects the rest to greater than average scarcity
>
> (Homer-Dixon, 1999a, p. 52)

Structural scarcity refers to the unequal access or distribution of natural resources (Bingham, 2001). This is a form of scarcity that applies only to certain groups that, relative to other groups, are excluded from equal access to particular resources. The unequal social distribution of a resource does not presuppose actual scarcity if the resource was distributed evenly (Urdal, 2008, p. 593). Van der Molen and Hildering (2005, p. 135) argued that structural scarcities emerge when more powerful segments of water users confiscate a larger part of the scarce resource.

This assumed that some individuals, groups, or states get disproportionately large slices of the pie while other groups get slices that are too small (Homer-Dixon, 1999a, p. 48). It implies a concentration of a resource in the hands of relatively few people while the remaining population suffers from serious shortages (i.e., unequal social distribution) (Percival and Homer-Dixon, 1998, p. 80). The perceived misallocation and unavailability of water have instigated clashes (Kreamer, 2012, p. 90). In addition, Homer-Dixon (1999a, p. 48) argued that this may cause violent conflicts (Homer-Dixon, 1999a, p. 8; Homer-Dixon and Blitt, 1998, p.).

Homer-Dixon (1999a, p. 48) asserted that structural scarcities arise from resources that are *excludable*, which means that property rights or other institutions can be used to prevent access to the resource by some actors. Renewable resources, such as river water have characteristics that permit the assignment of clear property rights. Conflicts over natural resources arise when parties disagree about the ownership, allocation, and use of natural resources (Homer-Dixon, 1999a, p. 48). The environmental scarcity theory notes that the increasing scarcity of renewable resources, or grievances over their governance, transboundary nature, or both can reinforce existing stress factors and contribute to the decision to resort to violence (UNEP, 2012, p. 14).

Resource scarcity is often caused by a severe imbalance in distribution. Such unequal distribution, or what is referred to as structural scarcity, is a key factor in virtually every case of scarcity that contributes to conflict. Often, the imbalance is deeply rooted in institutions (Homer-Dixon, 1999a, p. 5; Purkitt, 2009, p. 62).

Although it can be assumed that enough water is available for current and future needs, water resources are unevenly distributed. Elhance (1999, p. 4) argued that an unequal distribution of fresh water does not in itself necessarily lead to acute interstate conflict, but *severe scarcities of an essential, non-substitutable, and shared resource* like fresh water may make states prone to conflict. The unequal water distribution has significant implications for society, often causing widespread acute human suffering and economic damage on a continent where agriculture, largely rain-fed, is the single most important driver of economic growth (UNEP, 2010, p. 15).

It has been reported that cooperative or conflictive events associated with internationally shared water bodies are generally concerned with the allocation and use of water resources. One of the key challenges is allocating shared water resources and their benefits between upstream and downstream countries. Unequal distribution of water affects the quantity of water entering or leaving riparian countries, aside from possible changes due to climate change. Upstream water use would be expected to lead to increased water scarcity in the downstream parts of the basin (Munia *et al*, 2016, pp. 1–2).

The notion that river water is likely to cause interstate resource wars is deeply rooted in the fact that river water flows from one area to another. As a result, one country's access can be affected by another's actions. The argument that the scarcity of renewable resources like water will lead to violent conflict suggests that the physical control of these resources will provoke interstate simple scarcity conflicts (Homer-Dixon, 1999a, p. 228).

This argument is supported by the simple scarcity conflict perspective. Simple scarcity conflicts are,

> conflicts over scarce renewable resources between states. They are particularly likely to break out over resources that are essential for human survival and can be physically seized or controlled like river water, fisheries and agricultural productive land
>
> (Homer-Dixon, 1991, p. 87)

According to Homer-Dixon (1999a, p. 228), the word *simple* is used to distinguish this type of conflict from others that include psychological and social processes. Conflicts that include psychological and social processes are those embedded in a context of long-standing religious, cultural or worldview differences and inequalities, and occur mostly within countries rather than between countries. These may include ethnic clashes that arise from social cleavages. For instance, leaders may engage in conflicts to preserve their identity as a leader (i.e., fear of losing face) and to encourage group cohesiveness (Libiszewski, 1992, p. 12).

Environmental scarcity causes simple scarcity conflicts and increases society's demands on the state while decreasing its ability to meet those demands (Percival and Homer-Dixon, 1998, p. 281). The simple scarcity conflict argument posits political disputes and violent conflicts between states as occurring when,

> states rationally calculate their interests in a situation where there is a fixed or shrinking pie of natural resources
>
> (Homer-Dixon, 1999a, p. 137)

This forces states to seize or claim ownership of a shared renewable resource (Lipschutz, 1989, p. 46).

Regime theory

Understanding the effective responses to environmental change and mediation in nascent conflicts in transboundary river basins requires an analysis of the effect, nature, role, and significance of regimes, institutions or both. This analysis should rather be questioned in the following manner.

1 Can regimes deal with, respond to, and mediate emerging water conflicts?
2 If yes, how can they mediate between a variety of power and interest variables in a region to produce related behaviour and outcomes to foster cooperation?
3 Can developing countries respond to environmental problems facing transboundary river basins effectively to avert negative environmental effects?

According to Wettestad (1999, p. 173) and Young (2011, p. 19854), instances of environmental change or problems would require functioning regimes.
Moreover, another factor that adds to the stability of international watersheds is institutional resiliency. Solomon (2014, pp. 89–90) asserts that,

> once cooperative water regimes have been established through treaties, they tend to be tremendously resilient over time, even between otherwise hostile riparians and even as conflict is waged over other issues.

In an attempt to effectively manage shared water resources and solve common problems, institutions governing water resources need to absorb changes, cope with major climatic challenges, and adapt to new conditions (Nava, 2017, p. 3).

An important factor for enhancing and addressing the management and distribution of transboundary river basins is the implementation of institutional support structures. Godana (1985, p. 264) posits that basin states can only gain through the creation of a comprehensive commission serving as an institutional vehicle for cooperation. Moreover, the equitable development of transboundary river basins requires either an international institutional structure or a negotiated multilateral treaty regime as a bargaining tool between upstream and downstream countries (Okoth-Owiro, 2004, p. 24).

Godana (1985, p. 264) argued that all-inclusive agreements representing the needs and interests of all riparian states are imperative to ensuring that basins develop fully. According to Kukk and Deese (1996, p. 22), river basin organisations are essential as a means for all riparian countries to voice and resolve water issues without resorting to force. Similarly, Hasenclever *et al*. (2000, p. 14) noted that regimes may serve as institutional frameworks that facilitate the arrangement of side payments to improve the relative performance of otherwise dissatisfied actors. Therefore, many regimes allow relatively disadvantaged states to voice their concerns about the skewed distribution of gains and to push for correction (Hasenclever *et al*, 2000, p. 14).

Without these organisations, political tensions and disputes along shared rivers tend to emerge (Kukk and Deese, 1996, p. 22). Depletion, degradation, political tensions, and disputes emerge in many transboundary basins due to a lack of capacity and capability in the mediating river basin organisation. The organisation is thus unable to ensure cooperation for the rational planning, conservation, and development of resources in the basin (Godana, 1985, p. 264).

Similarly, the emergence of environmental issues in transboundary river basins is due to unilateral action, which is often ineffective and inefficient. As a result, transboundary cooperation is necessary and is shaped by, and contributes to, the development of transboundary management regimes (Raadgever *et al.*, 2008, p. 3).

Okoth-Owiro (2004, p. 25) believed that establishing an international river basin organisation, authority, or commission presents itself as the best solution for preventing and resolving water conflicts. They argued that a river basin organisation is needed to engage water scarce and rich countries in negotiations. Benvenisti (1996, p. 400) asserted that this allows for collective action so that all riparian countries enjoy access to resources that cross political boundaries.

A prominent factor that determines whether countries and their agricultural systems can respond to, adapt, or both effectively to the rising scarcities of land and water, are institutions. This simply means that if institutions fail, water will not be adequately preserved and agricultural systems will remain grossly inefficient (Homer-Dixon, 1999a, p. 87).

Since the 1960s, a growing academic interest in international regimes has culminated in extensive research into regimes, especially on international institutions and their capacity (Young, 1986, p. 104). This section focuses on regime formation, conceptualisation, relevance, and effectiveness.

A question often asked is: Can developing countries respond to environmental problems facing transboundary river basins effectively enough to avert these negative environmental effects? Problems within international river basin management have been explained mostly in descriptive ways. This section aims to provide a research framework that is based on a theoretically guided perspective.

According to Homer-Dixon (1999a, pp. 4–5), developing countries will feel the effects and impact of environmental scarcity more than developed countries. Developing countries' well-being is dependent on environmental goods and services, however, these countries have little, if any, financial, material, and human capital resources to prevent or adapt to environmental scarcities. This leaves their economic and political institutions fragile. The severe stress caused by environmental scarcity can constrain relations between states if the renewable resource in question is a shared resource.

Researchers argue that poor countries are particularly susceptible to resource conflicts as they cannot often adapt to environmental change. Societies have very different political, financial, and administrative capacities to respond adequately to increasing resource demands and such stresses can threaten stability and security. Substantial references to this debate appear in Homer-Dixon (1991; 1994; 1999a and b) and Homer-Dixon and Blitt (1998).

Some experts proposed that an interstate war may occur as a result of the shift in the balance of power between states, either regionally or globally, as a result of environmental change (Homer-Dixon, 1991, p. 77). Regime-based interstate cooperation is a reality that needs explanation. Africa had established regimes and institutions to achieve cooperation, with river-sharing agreements increasing in the 1990s. However, these agreements are presently under severe stress due to the increasing demand and decreasing supply of water resources. Moreover, the threat of global climate change has raised serious doubts about the future of these agreements (Swain and Krampe, 2011, p. 17).

The politics of transboundary water negotiations between states are complex as a result of the relations among riparian countries in that region (Munia *et al*, 2016, p. 2). The loss of water supplies physically manifests as local and regional problems. Water scarcity by itself does not trigger war; the issue is the governance of water. In conflict-prone countries that face poverty, livelihood pressures, and unaccountable government systems, an absence of clear water management policies in the face of reduced rainfall may overwhelm the government's ability to provide for the basic needs of its community (Abebe, 2009, p. 124).

Such governance failure can contribute to unrest and conflict, particularly in communities where low-intensity conflicts that are usually managed using local systems become complicated and fall beyond the realm of these coping mechanisms. This is especially true when they involve state-sponsored development projects, such as dam construction, and foreign investment in large-scale agriculture. The involvement of these actors will increase conflict intensity, which can lead to violence (Abebe, 2009, p. 124).

On international or interstate levels, environmental stress may trigger international disputes related to interstate conflicts on the access and control of water from rivers, and on obligations under multilateral environmental regimes (Tanzler *et al.*, 2002, p. 61). Resource depletion and degradation can influence ideational factors, such as prompting or impeding institutional reform. Ideational factors, for instance, institutions help to determine the kind of initiatives pursued in each region or society. They also help determine a region, society's, or both vulnerability and adaptability when faced with serious environmental scarcity (Homer-Dixon, 1999a, pp. 49–51).

Without a full understanding of regimes and institutions, analysts cannot begin to grasp the nature of the relationship in a given region between human activity and the scarcity of renewable resources, or why regions respond well or badly to this scarcity (Homer-Dixon, 1999a, pp. 49–51).

According to Hasenclever *et al.* (2000, p.7), cooperation makes all participants in a given region better off; however, it is hard to achieve the desired cooperation or collective action given the widespread uncertainty that characterises international life.

Regimes can and do reduce this uncertainty among states; they mitigate the fear of cheating or being exploited by the other parties; therefore, they make it easier for states to embark on collaborative ventures. In another approach, regimes reduce vulnerability and threats that are caused by the environment through building capacity in communities and governments to resolve environment-related conflicts. This can be achieved by providing funding, technical expertise, or support to address the threat directly (Hasenclever *et al.*, 2000, p. 7).

According to Neumayer (2001, p. 122), regime theory can address the question of how cooperation can be achieved and sustained in a world divided into sovereign nation-states. In this regard, regime theory is crucial when explaining how regimes create bargaining forums. It presupposes that states are governed by rules and norms (Neumayer, 2001, pp. 130–136). Regime theory assumes that when regimes, institutions, or both are in place, they can mediate between a variety of power and interest variables in a region to produce related behaviour and outcomes in an attempt to foster cooperation in collective security, among other things (Krasner, 1982, pp. 189–90).

Regime analysis focuses on patterns of interaction between states. It attempts to explain these relations in terms of principles, norms, rules, and decision-making procedures that arise and become stable because of the shared interests of the participants. Regime theory proposes that basic causal variables, including egoistic self-interest, political power, norms and principles, usage and custom, and knowledge, give rise to regimes (Thompson, 1991, p. 57).

Regimes consist of various types of institutions. A useful distinction can be made between formal and informal institutions. Formal institutions include law, policy, and government bodies, and are explicit and officially announced. Formal institutions that are important for transboundary river basin management are international and national law, policy, and governmental organisations. Law and policy make up the formal institutional environment, whereas governmental organisations are formal institutional arrangements. The institutional environment consists of fundamental political, social, and legal rules, and the institutional arrangements provide an organisational structure within which the members of society, individually or collectively, cooperate or compete (Raadgever, 2005, pp. 1–4).

Informal institutions include nongovernmental actors and cultural norms and values. However, these informal institutions are not explicit and officially announced; rather, they are followed or used in practice. Formal and informal institutions are relatively stable and durable features. In most sectors informal institutions are more durable; however, in the water sector, formal institutions change more slowly. Formal and informal institutions together influence operational water management and the performance of the water sector. The main factor that influences the performance of the water sector is institutional performance. Therefore, institutional settings or regimes are an important aspect of transboundary river basin management (Raadgever, 2005, pp. 1–4).

North (1990, p. 3) asserted,

> institutions are the formal rules of the game and are humanly devised constraints that shape interaction.

Institutions are also seen as,

> practices composed of recognised roles coupled with sets of rules or conventions governing relations among the occupants of these roles
> (Young, 1986, p. 108)

Institutions have, to varying degrees, rules for monitoring behaviour. By institutions, neoliberalists mean shared habits and practices of cooperation on the one hand, and fully developed systems of governance on the other (Keohane, 2012, p. 126).

Institutions evolve from the accumulated collective knowledge and are used as a substitute for information. In a world of perfect knowledge, institutions would not be necessary. However, in the absence of accurate information, institutions provide a basis for making reasonably sound decisions by regulating the behaviour of others (Raadgever, 2005, p. 3).

Liberal academics projected institutions as a powerful force for stability and order (Keohane, 1993, p. 53). Keohane (2012, pp. 125–126) asserted,

> Institutions and rules can facilitate mutually beneficial cooperation – within and among states.

However, according to Raadgever (2005, p. 3), a common definition of institutions is,

> rules or regularities of behaviour that are generally accepted by members of a social group, that specify behaviour in specific situations, and that are either self-policed or policed by external authority.

Keohane (2012, p. 126) insisted,

> The social purpose of Institutional Liberalism is to promote beneficial effects on human security, human welfare and human liberty as a result of a more peaceful, prosperous and free world. Institutional Liberalism justifies the use of power in constructing institutions on the basis of this conception of social purpose.

According to most liberal scholars[8], institutions play the main mediating role and act as the principal means to achieve and maintain cooperation between states. Mutual interests minimise differences between states and pave the way for cooperation. States become willing to cooperate when institutions (e.g., sets of rules and practices that prescribe roles, constrain activity, and shape the expectations of actors) are seen as beneficial. These theorists believe that institutions can and do have an impact on state behaviour.

The similarities, relationships, or both between regimes and institutions are illustrated by Young (1986) in the article *International Regimes*, where he claimed that regimes provide states with roles and responsibilities in a given region. They further assert that in understanding the machinery of regimes with regard to roles and responsibilities, international regimes should be viewed as social institutions. According to Young (1986, p. 107), social institutions are,

> recognised practices consisting of easily identifiable roles, coupled with collections of rules or conventions governing relations among the occupants of these roles.

They made a comparison to the roles and responsibilities of husband and wife in their institution of marriage (Young, 1986, p. 107).

According to Young (1986, p. 107), the rules and conventions that grow up around these roles and that constitute the superstructure of social institutions ordinarily encompass sets of rights or entitlements, as well as sets of behavioural prescriptions. In addition, it is not difficult to apply this conception of social institutions to the particular case of international regimes (Young, 1986, p. 107). Similar to regimes, it is evident that institutions matter in the conduct of state behaviour (Keohane and Martin, 1995, p. 40).

Regimes are created and maintained by states for the functions they serve in mixed-motive situations. Neoliberals point out that although states lack a sense of obligation, they think twice before they violate agreed rules. Moreover, since international institutions are difficult to construct, states are hesitant to put existing regimes at risk. This makes international regimes resilient, because they are economic and political investments (Hasenclever *et al.*, 2000, p. 8).

International environmental regimes may contribute to the emergence of an international framework that supports a peaceful settlement of conflicts. Countries that have practised cooperation on a number of issue areas should be able to resolve problems peacefully in other areas as well. However, Oberthur (2001, p. 244) also cautioned that international environmental policy has only in the rarest cases arrived at a sustained solution to underlying environmental problems. This is due to its structural limits, including the degree of effectiveness and the lack of effective sanctions to counter noncompliance. More effective international environment policies depend less on the specific organisational arrangements and primarily on the political will of countries to grant sovereignty to international bodies and to fully implement the agreements they adhere to domestically (Tanzler *et al.*, 2002, pp. 63–64).

Snidal (1996, p. 127) and Litfin (1997, p. 181 argued that when international regimes come into existence, they change actors' expectations and the rules of the game, make actors' decisions transparent, initiate further steps, and push issues forward. As a result, they start to constrain state behaviour. After regimes have been established, the states that created the regimes are no longer free to do as they please. States become accustomed to cooperation and successful interaction in one regime makes establishing the next one more likely. Snidal (1996, p. 127) and Litfin (1997, p. 181) maintained that successful regime building induces participating states to develop a collective identity that helps them to sustain cooperation, even in situations where either state would otherwise default.

Regional integration provides a context for peace building but can be a catalyst for conflict. This depends on the capacity, failure, or both of regimes to mediate differing interests over shared resources (Kameri-Mbote, 2005, p. 7), especially where states belong to several or different regimes because, in most instances, they pursue different agendas. Kameri-Mbote (2005, p. 7) highlighted the example of the countries in the Great Lakes Region, which belong to different economic integration bodies. These include the East African Community, Common Market for Eastern and Southern Africa, and South African Development Community (SADC).

Some of these bodies have protocols tailored specifically for environmental factors. However, these regional bodies overlap, and it is not unusual for neighbouring countries to be members of different regional organisations. For example, the DRC, Tanzania, and Zambia are members of SADC; however, Rwanda, Burundi, and Uganda are not. Further, Tanzania, Uganda, and Kenya are members of the East African Community, and the DRC, Rwanda, and Burundi are not. This assortment can affect or delay the effectiveness of regional bodies when building peace (Kameri-Mbote, 2005, p. 7).

Nonetheless, regimes have a responsibility to ensure the overall well-being of the environment. Given that many conflicts have the environment as a source of conflict, regimes must take an active role where environmental resources are related to the conflict. Some regimes are successful in addressing environmental conflicts by intervening directly at the threat level. In other words, if water scarcity is identified as the key conflict driver, regimes may employ a number of strategies to address this threat, for instance, by developing an agreement solely based on addressing water scarcity in a given region. Therefore, cooperation over shared water resources can be a focus around river basins.

On an international level, many multilateral international efforts by environmental regimes and international organisations (e.g., UN, UNEP, and the United Nations Development Programme [UNDP]) and international financial organisations and funds (e.g., World Bank and Global Environmental Facility) have been established to prevent environmentally-induced conflicts that may have a general effect of civilising and stabilising cooperation (Tanzler, *et al.*, 2002, p. 63).

There is a remarkable agreement concerning the reasons for actors in the international community being drawn into regimes or social institutions (Young, 1986, p. 109). Local tensions and violence, and increased regional tensions can, and do, occur over water issues. However, these tensions will likely continue to increase because,

> competition for water exists at all levels and is forecast to increase with demands for water in almost all countries
>
> (Gendron and Hoffman, 2009, p. 1)

As a result, a formation that governs water regimes in specific regions is required.

On a transboundary level, where a water supply is shared between two or more states, water regimes become an effective tool to promote cooperation over water resources. Such regimes shift the debate away from pure water sharing (e.g., natural versus historic rights) to a benefit-sharing approach (Earle, 2005, p. 64). According to Al-Din Amer (1997, p. 381), absolute territorial integrity (historic rights) protect(s) the rights of use for the country that puts the water into use first; therefore, giving downstream countries the power to accuse upstream countries of illegally taking measures disadvantageous to their interests. However, absolute territorial sovereignty (natural rights) regards water bodies as an integral part of the state's national territory. This principle protects the sovereignty of countries within which watersheds, lakes, and rivers originate.

The demand for regime development is also influenced by factors including the actors' deliberate agreement to institutional arrangements through some sort of bargaining process, as long as the marginal benefits of doing so outweigh the marginal costs (Young, 1986, pp. 109–110).

Why regimes are formed is self-evident. According to Lindemann (2009, p. 701), a water regime is formed when riparian countries engage in norm and rule-based cooperation for the political resolution of political problems and conflicts in the field of international river basin management. The main purpose of water regimes is to provide environmental governance in a given region. The conditions for the success of international water regimes involve two interrelated, but analytically separated, questions that focus on the determinants of water regime formation and water regime effectiveness. First, under which conditions are riparian states prepared to form institutions to address transboundary problems? Second, what are the determinants of the effectiveness of those institutions?

The question of regime effectiveness depends on numerous issues. For instance, regime effectiveness centres on the degree to which a regime eliminates or alleviates the problem that prompts its creation (Young, 1999, p. 4). Lindemann (2009, p. 701) further argued that regime effectiveness depends on behavioural change. Regime effectiveness becomes the extent to which the water regime prompts behavioural change on the part of relevant actors, leading to improved management of the problem.

Lindemann (2009, p. 701) identified five determinants of water regime formation and effectiveness. These are: problem; process; institutional; country-specific; and international context factors.

Problem factors

Marty (2001, pp. 35–36) distinguished between problems related to transboundary externalities and collective problems. Transboundary externality problems arise due to the upstream–downstream structure of international rivers. These externalities can be negative and positive.

The negative is a result of the costs imposed by the upstream country on the downstream country without compensation for the harm inflicted, such as in the case of water abstraction or pollution upstream. The positive occurs less frequently. These are situations where one riparian country produces a public good without receiving full compensation for its efforts, such as providing flood control upstream.

Collective problems, such as floods or common development projects, are caused, affect or both all riparian countries as a collective (Marty, 2001, p. 37).

Process factors

Forming water regimes generally requires the development of instruments to reduce the perceived costs of regime formation. There are two types of costs in this regard: information costs; and negotiation costs.

Information costs are related to uncertainty about the nature of the respective problem and the behaviour of the other parties, third-party actors, or both. Negotiation costs result from communication problems, fragmented decision-making procedures, or the high number of actors involved. As such, political tools such as confidence-building meetings at the technical level or the exchange of data can help to minimise information and negotiation costs (Lindemann, 2009, p. 702).

Institutional factors

Successful water regimes need to be specific, feasible, flexible, equipped with a centralised organisational structure, and open. Specific regimes are problem-oriented and incorporate precise rules and procedures that structure the relevant actors' behaviour to better manage the problem. Without these precise rules and procedures, there is scope for interpretation and rule avoidance, which will have a major impact on the effectiveness of the regime (Lindemann, 2009, p. 702).

Feasible regimes choose their goals according to available financial and personal resources. Where the goals outweigh the available financial and personal resources, the effectiveness of the regime is expected to decrease. Given that resources to solve the problems related to international river basin management are usually limited, there are reasons to assume that water regimes tend to be more effective if their goals are of limited scope (Marty, 2001, p. 47).

Flexible regimes include institutional mechanisms that allow for adaptation to changes in the problem structure. A lack of regime flexibility is likely to lead to reduced regime effectiveness since the existing problem-solving strategy may prove to be inadequate to cope with the changing circumstances.

For water regimes to be effective, they also need a centralised organisation structure. This will enhance and facilitate coordination, communication, and monitoring that will, in turn, lead to a more effective regime (Lindemann, 2009, p. 702).

Open regimes encourage a high degree of public participation. Where nonstate actors and civil society are involved in decision-making, the available knowledge and legitimacy of the respective regime should increase and, thus, foster its effectiveness (Lindemann, 2009, p. 702).

Country-specific factors

According to Mitchell (2001, p. 13),

> Determinants of international regime effectiveness are not restricted to institutional design but also include country specific characteristics.

This involves the specific features of regime member countries that influence the implementation and success of regimes (Brown-Weiss and Jacobson, 1998, p. 7). A specific and important feature of regime member countries is capacity (Lindemann, 2009, p. 703).

The capacity of regime member countries to implement the agreed standards, impose behavioural, or both on relevant national actors is very different. Where these capacities are insufficient, the effectiveness of the regime is impaired. Two important and influential capacities of regime member countries that will affect the prospects of a regime are economic–technological capacities and political–institutional capabilities (Lindemann, 2009, p. 703).

If regime members do not possess sufficient financial and administrative resources to plan and administer international water projects and the technical capability to generate data and implement projects at a national level, then water regime effectiveness is likely to suffer. Similarly, if the regime member countries are characterised by high levels of political stability, the prospects are good for high water regime effectiveness (Lindemann, 2009, p. 703).

Factors of international context

According to Mitchell (2001, p. 12), factors of international context can influence the effectiveness of international regimes. This relates to the bilateral relationships between regime member countries because of historical and current interactions and conditions, patterns of communication, and levels of trust. If bilateral relationships are strained by historical or current foreign policy conflict, there is no basis of trust to implement, further develop, or both the water regime. Its effectiveness is therefore bound to suffer.

Furthermore, Mitchell (2001, p. 13) asserted that the effectiveness of international regimes varies with the influence of international organisations. International organisations provide financial resources, offer their technical and cognitive expertise, and act as independent mediators between conflict parties (Mitchell, 2001, p. 13). International organisations provide for regime effectiveness, especially in developing countries where financial, technical, and cognitive resources are scarce (Brown-Weiss and Jacobson, 1998, p. 7).

Conclusion

Homer-Dixon's environmental scarcity theory adequately guides future research into the relationship between environmental scarcity and conflict through its tripartite definition and approach to environmental scarcity. Therefore, population growth, in combination with consumption levels, makes environmental degradation and depletion worse (Homer-Dixon, 1999a, p. 55).

The fundamental basis of the environment and conflict linkages explained that resource scarcity, as a result of population growth, environmental degradation, and resource distribution, communicates a clear and common message to a wide variety of conflict specialists (Kameri-Mbote, 2005, p. 2).

Understanding the link between environmental scarcity and conflict requires the analysis of the effects and nature of environmental scarcity.

The security implications of environmental scarcity are grave. A key challenge today is to better understand the relationship between environmental degradation (e.g., caused by rapid population growth, land degradation, climate change, migration, and unequal distribution of resources) and conflict (Idule-Amoko, 2009, p. 110). Environment–conflict linkages require scrutiny (Goulden and Few 2011, p. 6). Environmental scarcity, if not well managed, will increasingly threaten humanity's shared interests and collective security in many parts of the world. It will also pose a variety of social, political, and strategic challenges to many regional organisations.

A growing population and other factors, such as a decrease in rainfall and increased use of water for industrial and agricultural production, are likely to contribute to an unequal relationship between water supply, demand, and distribution (Pimentel *et al.*, 1996). Dissociating one element of environmental scarcity causing a change in environmental, social, economic, and political realities from other multiple elements that shape and influence regional transboundary basins, remains a challenge. This reflects the underlying analysis that one environmental feature has an impact in combination with other features of the environmental, social, economic, and political landscape.

The increasingly negative consequences of the depletion and degradation of land and water resources have influenced academics and governments worldwide to acknowledge the need to minimise the phenomenon and restore degraded land and watersheds (Van Schaik and Dinnissen, 2014, p. 11). With or without these challenges, Africa must use and share its transboundary resources collectively, equitably, and profitably for the benefit of its present and future social, economic, and political needs. It is only through such cooperative efforts that cooperation can be enhanced and interstate tension and conflict avoided.

One issue that was quite evident regarding the effectiveness and formation of regimes was that problem and process factors determine the prospects of regime formation and the last three factors (e.g., institutional, country specific, and international context factors) determine the effectiveness of regimes.

An important negative social effect of environmental scarcity that regimes should address is the constraint that environmental scarcity imposes on regional economic development, such as domestic, agricultural, and industrial production (Homer-Dixon, 1999a, p. 88).

Institutions, regimes, agreements, laws, and regulations for efficiently and effectively distributing environmental goods and services among states are imperative.

The analysis of resources and conflicts requires a thorough understanding of institutions that shape the rules and rights of resource use. Different layers of environmental governance at local, national, and international levels need to be incorporated into the analysis of resource use conflicts. As such, when determining whether regimes, institutions, or both matter, researchers need to analyse their significance as determinants of collective behaviour at the international level (Young, 1986, p. 115).

The most important factor in ensuring the long-term sustainable development of water resources remains in the realm of social, economic, and institutional development. The focus of cooperation should be on sharing the benefits of the water resource, rather than strict adherence to sharing the water itself. A move from the top-down management approach towards a participatory system, which links communities that share a resource, is needed. These communities need to be resilient to the drivers of change, such as those brought about through changes in climate and population, by developing networks with other communities. The focus should be on trying to upscale local cooperation over shared water resources at the national and regional levels.

Whether scarcity problems are resolved peacefully or not depends largely on the effectiveness of governing institutions. Environmental stresses are particularly aggravated when combined with rapid population growth. As the regime weakens, its ability to manage emerging conflict becomes more limited.

Effective climate and environmental policies hold the promise of reducing such potential and serve as tools for conflict prevention and conflict avoidance (Tanzler *et al.*, 2002, p. 4). Keohane (1993, p. 53) asserted,

> avoiding conflict depends greatly on institutionalized cooperation.

To date, many conflicts on competing claims for transboundary water resources (e.g., rivers and aquifers) have been solved by negotiations (hydro-diplomacy) and international agreements (e.g.., bilateral and multilateral agreements) (Tanzler *et al.*, 2002, p. 63; Mahlakeng, 2020).

Good governance is likely to benefit land management practices, food and water policies, and economic growth, as well as prevent tension and conflict. Governance is important, because it defines who has the power to make decisions about the ownership, consumption, and distribution of resources. Equally so, dysfunctional institutions, or bad governance practices, are considered strong indicators that conflicts will emerge and environmental stresses will worsen.

The majority of existing institutions governing international rivers are increasingly turning volatile due to the increased demand and decreased supply of fresh water. Adding to the problem is the threat of climate change, which has started to undermine the ongoing regimes and institutions relating to water sharing and the management of international rivers (Swain and Krampe, 2011, p. 17).

According to Idule-Amoko (2009, p. 111), Africa lacks policy, legal, and institutional frameworks within national, regional, and continent-wide mechanisms to develop transboundary resources (Idule-Amoko, 2009, p. 111). Authors such as Swain and Krampe (2011, p. 21) argued that the existing water-sharing regimes in Africa cannot address the emerging challenges that climate change will pose. Most of the ongoing cooperation in international river basins in Africa originated from the active involvement of international donor agencies. Many of these initiatives only barely survive because of external aid and assistance. Such reliance on minimal external help, alone, exposes the lack of interest, capacity, or both of basin states in Africa when creating effective and sustainable management for shared river resources.

Such authors further argue that to address the imminent problems of the existing water crisis and the massive change in the runoff structure due to climate change, the ownership and, most importantly, the accountability of transboundary water management must be restored to the countries in the regions. The role of the international community should be limited as far as regional water-sharing politics is concerned. This will enable and encourage riparian countries to find distinct emancipatory approaches to basin-based river management. These approaches should address the regions' unique cultures and history, as well as their economic disparity and ecological sensitivities (Swain and Krampe, 2011, p. 21).

To sustain and develop a resource that satisfies a basic condition of life will be the biggest challenge for basin states and Africa's transboundary river management processes. The key to transboundary river management for cooperation rather than confrontation will be the smart governance of natural resources, and how politics deals with water issues and emerging threats (Swain and Krampe, 2011, p. 21). Regimes that focus on addressing vulnerabilities and threats that are caused by the environment may prevent or mitigate conflicts. Given the complexity of conflicts, regimes should avoid focusing on one element of environmental insecurity.

An aspect that features in the arguments of Homer-Dixon's environmental scarcity and regime theories is that the environmental scarcity of renewable resources can indirectly help generate instability, whereas institutions, due to their capacity, often appear to determine cooperation or conflict (Homer-Dixon and Blitt, 1998, p. 223). It is evident that a large number of factors (e.g., human and natural causes, and institutions) potentially influence the emergence and continuation of conflicts, and that they are usually interrelated.

In conclusion, the environmental quantity and quality of a country or region can be causally linked to the presence or absence of conflict. It is in this manner that we can safely and adequately merge Homer-Dixon's environmental scarcity and regime theories within the context of transboundary river basins.

Notes

1 led by Thomas Homer-Dixon.
2 led by Günther Bächler and other scholars from the Swiss Peace Foundation and the Centre for Security Studies and Conflict research.
3 'people who are forced to leave their home region due to sudden or long-term changes to their local environment which compromise their well-being or secure livelihood,

such changes are held to include increased droughts, desertification, sea level rise, and disruption of seasonal weather patterns' (Terminski, 2011:1).

4 Norman Myers, (1996) *Ultimate Security: The environmental basis of political stability* and (2004) *Environmental security: What's new and different?*

5 'A change in the weather pattern that directly results from the increased warming temperatures of the sea at certain points of the Pacific Ocean (Equatorial Pacific)' (L'Heureux, 2014).

6 See Chapter 4

7 'Any physical, chemical, and/or biological changes in the ecosystem (or resource base) that render it, temporarily or permanently, unsuitable to support human life' (El-Hinnawi, 1984, p. 4).

8 Keohane and Nye, 1977; Keohane, 1984, 1989; Axelrod and Keohane, 1985; Haas, Keohane and Levy, 1993; Lipson, 1984; Milner, 1992

Bibliography

Abdi, O.A., Glover, E.K., and Luukkanen, O. (2013) 'Causes and impacts of land degradation and desertification: Case study of the Sudan', *International Journal of Agriculture and Forestry*, 3(2).

Abebe, L. (2009) 'Natural resource conflicts in West Africa: The case of the Niger River Basin', in D.W. Tadese (ed.), *Climate change and transboundary water resource conflict in Africa*. Mombasa Institute for Security Studies (ISS).

Ahmed, F. (2011) 'Key social vulnerabilities to climate change in South Africa's coastal zones: The potential for conflict', in V. Gounden (ed.), *Environment, Climate Change and Conflict*. *Conflict Trends*, Issue 2. Durban: The African Centre for the Constructive Resolution of Disputes (ACCORD).

Al-Din Amer, S. (1997) 'The law of water: Historical record', in Dupuy B. (ed.) *Aspects économiques de la gestion de l'eaudans le bassinméditerranéen*. Bari: Centre International de Hautes Etudes Agronomiques Méditerranéennes.

Axelrod, R., and Keohane, R.O. (1985) 'Achieving cooperation under anarchy: Strategies and institutions', *World Politics*, 38(1).

Baechler, G. (1998) 'Why environmental transformation causes violence: a synthesis', Environmental change and security program (ECSP), Project report 4. Washington DC: The Wilson Center.

Baecher, G. *et al.* (2000) 'The Nile Basin: Environmental transboundary opportunities and constraints analysis', Washington, DC: United States Agency for International Development (USAID) Draft Paper.

Benvenisti, E.W. (1996) 'Collective action in the utilization of shared freshwaters: The challenges of international water resources law', *American Journal of International Law*, 90.

Barnett, J. (2000) 'Destabilizing the environment – Conflict thesis', *Review of International Studies*, 26(2).

Benjaminsen, T.A. (2008) 'Does supply-induced scarcity drive violent conflicts in the African Sahel? The case of the Taureg rebellion in Northern Mali', *Journal of Peace Research*, 45(6).

Bingham, K. (2001) 'Environmental scarcity and the outbreak of conflict', *Population Reference Bureau Information*. Available at: www.prb.org/Publications/Articles/2001/EnvironmentalScarcityandtheOutbreakofConflict.aspx. (Accessed: 29 December 2013).

Bronkhorst, S. (2011) 'Addressing climate-related conflict: Human security and lessons from the southern Sahelian belt of Sudan', in Gounden, V. (ed.) *Environment, climate change and conflict, conflict tends*, Issue 2. Durban: ACCORD.

Brown-Weiss, E., and Jacobson, H.K. (1998) *Engaging countries: Strengthening compliance with international environmental accords*. Cambridge, Mass: MIT Press.

Conway, D. *et al.* (2009) 'Rainfall and water resources variability in sub-Saharan Africa during the twentieth century', *Journal of Hydrometeorology*, 10.

Diehl, P.F., and Gleditsch, N.P. (2001) *Environmental conflict: An anthology*. Boulder, CO: Westview Press.

Dieperink, C. (1998) 'From open sewer to salmon run: Lessons from the Rhine water quality regime', *Water policy*, 1(5).

Di Nunzio, J. (2013) 'Conflict on the Nile: The future of transboundary water disputes over the world's longest river',. *Future Directions International*.

Earle, A. (2005) 'Hydropolitics in Southern Africa: What is the prognosis for peaceful development of shared watercourses?', Forschungsinstitut Wasser und Umwelt (FWU)/ University of Siegen Research Institute for Water and Environment FWU, Vol.3, Topics of integrated Watershed Management – Proceedings, Summer School. University of Siegen Research Institute for Water and Environment. Siegen.

Elhance A. (2000) 'Hydropolitics: Grounds for despair, reasons for hope', *International Negotiation*, 5(2).

El-Hinnawi, E. (1984) *Environmental refugees*. Nairobi, Kenya: UN Environmental.

Gendron, R. and Hoffman, E. (2009) 'Resource scarcity and the prevention of violent conflicts', *Peace & Conflict Review*, 4(1).

George, B.H. and Cowie, A.C. (2011). 'Bioenergy systems, soil, health and climate change', in Sign, B.P. Cowie, A.L. and Chan, K.Y. (eds.) *Soil, health and climate change*. Berlin, Heidelberg: Springer,

German Environment Ministry. (2002) 'Climate change and conflict: Can climate change impacts increase conflict potentials? What is the relevance of this issue for the international process on climate change?'. Berlin: Federal Ministry for the Environment, Nature Conservation and Nuclear Safety.

Gleditsch, N.P. (1997) *Environmental conflict and the democratic peace*. Oslo. Kluwer Academic Publishers.

Gleick, P. (1989) 'Climate change and international politics: Problems facing developing countries', *Ambio*, 18(6).

Godana, B.A. (1985) *Africa's shared water resources*. London: Frances Pinter, Boulder, Colorado: Lynne Rienner Publishers.

Goulden, M. and Few, R. (2011) 'Climate change, water and conflict in the Niger River Basin', Washington, DC: USAID.

Haas, P. M., Keohane, R. O., and Levy, M. A. (1993). *Institutions for the Earth: Sources of effective international environmental protection*. Cambridge, MA: MIT Press.

Hagmann, T. (2005) 'Confronting the concept of environmentally induced conflict', *Peace, Conflict and Development*, 6.

Hasenclever, A., Mayer, P. and Rittberger, V. (2000) 'Integrating theories of international regimes', *Review of International Studies*, 26.

Hendrix, C.S., and Salehyan, I. (2010.) 'After the rain: Rainfall variability, hydrometeorological disasters, and social conflict in Africa', 52nd *International Studies Association Annual Conference, Global Governance: Political Authority in Transition*. Montreal, 16–19 March 2011.

Hernandez, N. E. (2012) 'Water security conflicts: A regional perspective', *Small Wars Journal*, Sep 28. Available at: https://smallwarsjournal.com/jrnl/art/water-security-conflicts-a-regional-perspective. (Accessed: 7 November 2022).

Homer-Dixon, T.F (1991) 'On the threshold: Environmental changes as causes of acute conflict', *International Security*, 16(2).

Homer-Dixon, T.F. (1994) 'Environmental scarcities and violent conflict: Evidence from cases', *International Security*, 19(1).

Homer-Dixon, T.F., and Blitt, J. (1998) *Ecoviolence: Links among environment, population, and security*. New York: Rowman and Littlefield Publishers.

Homer-Dixon, T.F. (1999a) *Environment, scarcity, and violence*. New Jersey:
Princeton University Press.

Homer-Dixon, T.F. (1999b). 'Threshold of turmoil: environmental scarcities and violent conflict', in Deudney D.H. and Matthew R.A. (eds.) *Contested grounds: Security and conflict in the new environmental politics*. Albany: State University of New York Press.

Idule-Amoko, A. (2009) 'The challenges of climate change and transboundary resources in Eastern Africa', in Tadese D.W. (ed.) *Climate change and transboundary water resource conflict in Africa*. Mombasa: Institute for Security Studies.

Kameri-Mbote, P. (2005) 'Environment and Conflict linkages in the Great Lake region', International Environmental Law Research Centre (IELRC) Working Paper.

Kennedy, B. (2001) 'Environmental scarcity and the outbreaks of conflict. Population Reference Bureau'. Available at: http://www.prb.org/Publications/Articles/2001/EnvironmentalScarcityandtheOutbreakofConflict.aspx. (Accessed: 24 October 2013).

Keohane, R.O. and Nye, J.S. (1977) *Power and interdependence*. Boston: Little, Brown.

Keohane, R.O. (1984). *After hegemony: Cooperation and discord in the world political economy*. Princeton, N.J.: Princeton University Press.

Keohane, R.O. (1989). *International institutions and state power, essays in international relations theory*. Boulder, CO: Westview Press.

Keohane, R.O. (1993) 'The diplomacy of structural change: Multilateral institutions and state strategies', in Haftendorn, H. and Tuschhoff, C. (eds.) *America and Europe in an era of change*. Boulder, Colorado: Westview Press.

Keohane, R.O. and Martin, L.L. (1995) 'The promise of institutionalist theory', *International Security*, 20(1).

Keohane, R.O. (2012) 'Twenty years of institutional liberalism', *International Relations*, 26(2).

Klare, M.T. (2001) 'The new geography of conflict', *Foreign Affairs*, 80(3).

Klare, M.T. (2013) 'Resource shock: How resource scarcity and climate change could produce a global explosion. Tom Dispatch', 21st April. Available at: www.alternet.org/environment/resource-shock-how-resource-scarcity-and-climate-change-could-produce-global-explosion. (Accessed: 9 August 2014).

Krasner, S. (1982) 'Structural causes and regime consequences: Regimes as intervening variables', *International Organisation*, 36(2).

Kreamer, D.K. (2012) 'The past, present, and future of water conflict and international security', *Journal of Contemporary Water Research and Education*, 149(11).

Kukk, C. and Deese, D. (1996) 'At the water's edge: Regional conflict and cooperation over fresh water', *UCLA Journal of International Law and Foreign Affairs*, 21(1).

Lam, A. (2012a) 'The rising tide – Environmental refugees', *The New America Media*, 15 August. Available at: http://newamericamedia.org/2012/08/the-rising-tide----environmental-refugees.php. (Accessed: 19 July 2016).

Lam, A. (2012b) 'Number of environmental refugees to grow to 50 million by 2020', *Earth Island Journal*. 23 August. Available at: www.earthisland.org/journal/index. php/elist/eListRead/number_of_environmental_refugees_to_grow_to_50_million_by_2020/. (Accessed: 19 July 2016).

Leroy, M. and Gebresenbet, F. (2011) 'Climate conflicts in the Horn of Africa'. in Gounden, V. (ed.), *Environment, Climate Change and Conflict, Conflict Trends*, Issue 2. Durban: ACCORD.

Libiszewski, S. (1992) 'What is an Environmental Conflict?', *first coordination meeting of Environment and Conflict Project*, Zurich. 30 April–1 May.

Lindemann, S. (2009) 'Success and failure of international river basin management: The case of Southern Africa', in Brauch, H.G. *et al.* (eds.) '*Facing global environmental change: Environmental, Human, Energy, Food, Health and Water Security concepts*', Hexagon Series on Human and Environmental Security and Peace, Vol .4. New York: SpringerLink.

Lipschutz, R. D. (1989). *When nations clash: Raw materials, ideology and foreign policy*. New York: Ballinger.

Lipson, C. (1984) 'International cooperation in economic and security affairs', *World Politics*, 37(1).

Litfin, K.T. (1997) 'Sovereignty in world ecopolitics', *Mershon International Studies*, 41(2).

Mahlakeng, M.K. (2020) 'Tensions on the Nile: A need for hydro-diplomacy in the shifting hydropolitics', *Research on Islam and Muslims in Africa (RIMA)*, 8(15). Available at: https://muslimsinafrica.wordpress.com/2020/11/10/tensions-on-the-nile-a-need-for-hydro-diplomacy-in-the-shifting-hydropolitics-dr-mahlakeng-khosi-mahlakeng/. (Accessed: 9 July 2021).

Marty, F. (2001) 'Managing international rivers: Problems, politics and institutions, European university studies: Series 31', *Political Science*, 421.

Milner, H. (1992) 'International theories of cooperation among nations: strengths and weaknesses', *World Politics*, 44(3).

Mitchell, R.B. (2001) 'Comparative analysis of the effectiveness of environmental treaties', (unpublished paper) University of Oregon.

Munia, H. *et al.* (2016) 'Water stress in global transboundary river basins: Significance of upstream water use on downstream stress', *Environmental Research Letters*, 11(1).

Myers, N. (1996) *Ultimate security: The environmental basis of political stability*. Washington, DC: Island Press.

Nava, L.F. (2017) 'Water security through institutional resilience in the transboundary Paso del Norte region', *World Environmental and Water Resources Congress 2017. International Perspectives, History and Heritage, Emerging Technologies, and Student Papers*. Sacramento, May 21–25.

Neumayer, E. (2001) 'How regime theory and economic theory of international environmental cooperation can learn from each other', *Global Environmental Politics*, 1(1).

Nordas, R. and Gleditsch, N.P. (2007) 'Climate change and conflict', *Political Geography*, 26.

North D. (1990). *Institutions, institutional change and economic performance*. Cambridge: Cambridge University Press.

Oberthur, S. (2001) 'Preventing environmentally-induced conflicts through international environmental policy, in Carius, A., and Vincze, A (eds.) *Responding to environmental conflicts: Implications for theory and practice*. Budapest: Kluwer Academic Publishers.

Ohlsson, L. (1999) *Environment, scarcity and conflict: A study of Malthusian concerns*. Göteburg: Department of Peace and Development Research.

Okoth-Owiro, A. (2004)'The Nile Treaty: State succession and International Treaty Commitments', A case study of the Nile water Treaties', Occasional Papers series Issue 9. Nairobi: Konrad Adenauer Foundation.

Percival, V. and Homer-Dixon, T. (1998) 'Environmental scarcity and violent conflict: The case of South Africa', *Journal of Peace Research*, 35(3).

Pimentel, D., Huang, X., Cordova, A. and Pimentel, M. (1996) *'Impact of population growth on food supplies and environment', American Association for the Advancement of Science (AAAS) Annual Meeting*, Baltimore, 9 February, Available at: www.dieoff.com/page57.htm. (Accessed: 7 July 2016).

Purkitt, H.E. (2009) *African environmental and human security in the 21st century*. Amherst, New York: Cambria Press.

Raadgever, T. (2005) *Transboundary river basin management – state-of-the-art review on transboundary regimes and information management in the context of adaptive management.* Deliverable 1.3.1 of the NeWater project. Delft: Centre for River Basin Administration, Delft University of Technology

Raleigh, C., and Kniveton, D. (2012) 'Come rain or shine: an analysis of conflict and climate variability in East Africa', *Journal of Peace Research*, 49(1).

Roudi-Fahimi, F., Creel, L., and De Souza, R. M. (2002) 'Finding the balance: Population and water scarcity in the Middle East and North Africa'. Washington, D.C: Population Reference Bureau.

Schwartz, D.M., Deligiannis, T., and Homer-Dixon, T.F. (2000) 'The Environment and violent conflict: A response to Gleditsch's critique and some suggestions for future research', *Environmental Change & Security Project Report*, Issue 6.

Simpkins, G. (2010) 'Africa faces water war threats', Africa Rising 21st Century. 26 February. Available at: http://africarising2010.blogspot.co.za/2010/02/africa-faces-water-war-threats.html. (Accessed: 11 March 2016).

Snidal, D. (1996) 'Political economy and international institutions', *International Review of Law and Economics*, 16(1).

Solomon, H. (2014) 'Potential for cooperation rather than conflict in the face of water degradation: The cases of the Nile River and Okavango River Basins', *Southern Journal for Contemporary History*, 39(1).

Swain, A. and Krampe, F. (2011) 'Transboundary rivers and climate change: African and Asian Rivers'. in Gounden, V. (ed.) *Environment, Climate Change and Conflict, Conflict Trends*, Issue 2. Durban: ACCORD.

Tadesse, D. (2010) 'The impact of climate change in Africa', *Institute for Security Studies*, 220(11).

Tadese, K.W. (2013) *International watercourses law in the Nile River Basin: Three States at a crossroads*. New York: Routledge.

Tanzler, D., Carius, A. and Oberthur, S. (2002) 'Climate change and conflict prevention: Can climate change impacts increase conflict potentials? What is the relevance of this issue for the international process on climate change?' *Environmental Policy*.

Terminski, B. (2011) 'Towards recognition and protection of forced environmental migrants in the public international law: Refugee or IDPs umbrella', Washington: Policy Studies Organization Summit Proceedings. "Science, Technology & Environmental Policy: Pressing Issues, Little Time", Washington D.C. 18 December.

Thompson, L. (1991) 'Of myths, monsters and money: Regime conceptualisation and theory in the Southern African context', *Journal of Contemporary African Studies*, 10(2).

Tsuma, W. (2011) 'Climate change-conflict nexus: Framework for policy-oriented action', in Gounden, V. (ed.) *Environment, climate change and conflict, conflict trends*, Issue 2. Durban: ACCORD.

UN. (2015) 'International Migration Report', *Economics and Social Affairs*. Available at: www.un.org/en/development/desa/population/migration/publications/migrationreport/docs/migrationreport2015_highlights.pdf. (Accessed: 20 July 2016).

UNEP. (2006) 'Environment, sustainable development and the Nile River Basin', Education for Sustainable Development Innovations. MESA University Partnership. United Nations Environment Programme (UNEP): Available at: https://docplayer.net/2630438-Education-for-sustainable-development-innovations-mesa-universities-partnership-environment-sustainable-development-and-the-nile-river-basin.html. (Accessed: 9 November 2017).

UNEP (2010) *Africa water atlas, Division of Early Warning and Assessment (DEWA), (UNEP)*: Nairobi: Progress Press Limited, Malta.

UNEP. (2012) *Renewable Resources and Conflict, Toolkit and Guidance for Preventive and Managing Land and Natural Resources Conflict, the United Nations Interagency Framework Team for Preventive Action*. New York: UNEP.

Urdal, H. (2008) 'Population, resources, and political violence: A subnational study of India, 1956-2002', *Journal of Conflict Resolution*, 52(4).

Van der Molen, I. and Hildering, A. (2005) 'Water: Cause for conflict or co-operation?' *Journal on Science and World Affairs*, 1.

Van Schaik, L. and Dinnissen, R. (2014) 'Terra incognita: Land degradation as underestimated threat amplifier', Wassenaar: Netherlands Institute of International Relations.

Weiner, M, and Russell, S. S. 2001. *Introduction Demography and national security*. New York and Oxford: Berghahn.

Werz, M. and Conley, L. (2012) *Climate change, migration, and conflict addressing complex crisis scenarios in the 21st Century*. Washington, D.C: Center for American Progress:

Wettestad, J. (1999) *Designing effective environmental regimes*. Northampton, MA: Edward Elgar Publishing Limited.

Williams, K.R. and Gemenne, F. (eds.) (2016) *Organisational perspective on environmental migration*. London: Routledge.

Young, O.R. (1986) 'International regimes: Toward a new theory of institutions', *World Politics*, 39(1).

Young, O.R. (2011) 'Effectiveness of international environmental regimes: Existing knowledge, cutting-edge themes, and research strategies' *Proceedings of the National Academy of Sciences of the United States of America*, 108(50).

3 The Orange Senqu River Basin

The Orange Senqu River Basin (OSRB) is characterised by a high economic dependence on local natural resources in the form of agriculture (frequently subsistence) and pastoralism, in which the variability in the climate and the availability of water to a large extent determine production. This dependence requires a lot of water use and coupled with unsustainable agricultural practices, often leads to depleted water resources.

The concept of integrated water resources management (IWRM) has been broadly endorsed as a framework for addressing water scarcity and conflict, and as a key to sustainably managing water resources (Kistin, 2010, p. 26). The IWRM seeks to promote the coordinated development and management of water, land, and related resources to maximise economic and social welfare equitably without compromising the sustainability of vital ecosystems (Kistin, 2010, p. 26). According to Hepworth (2009, p. 19), IWRM represents,

> a discursive response to the knowledge that the interconnectedness of the water cycle, particularly at river catchment and aquifer scale, requires that the full range of water uses and their needs be considered together to avoid conflicts and negative impacts and to optimize the benefits of water use.

When analysing the state of hydropolitical vulnerability in southern Africa, Earle (2005, p. 56) suggested that it is essential to look at both sides of the equation, for example, the underlying factors that create vulnerability to conflict as a response to hydrologic change as well as the legal and institutional response to such change.

Description of the OSRB

The OSRB is located in southern Africa and extends over four countries (Langea *et al.*, 2007, p. 660). These include the Republic of Botswana, the Kingdom of Lesotho, the Republic of Namibia, and the Republic of South Africa. The basin rises in the Kingdom of Lesotho and flows westwards for 2,300 km into the Atlantic Ocean (*Figure 3.1*). The OSRB faces seasonal and spatial water scarcity (Meyer, 2013, p. 28).

DOI: 10.4324/9781003366379-3

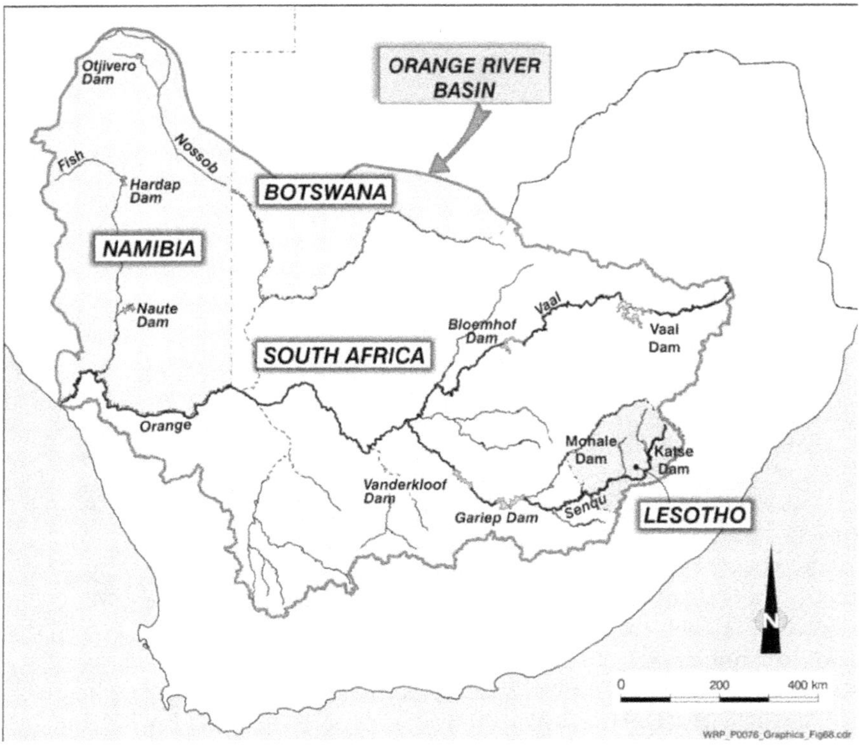

Figure 3.1 The OSRB.

Of the river's total flow of 11,300 million/m^3 per year, South Africa withdraws about 63%, making it the largest water user among the Orange River riparian states. Mostly relying on groundwater, downstream Namibia uses only 1.3% of the water and upstream Lesotho, with its plentiful water supply, withdraws a mere 0.2%.

While Botswana is considered an Orange River riparian country due to its hydrological links to the ephemeral Molopo–Nossob tributary, at present it does not use the Orange River surface waters, as there are few inhabitants and little industrial development in its share of the basin (Sullivan, 2014, p. 189)

According to the South African Department of Water and Sanitation (DWS) (2016),

> The natural resources of the OSRB are estimated to be in order of 12 000 million m^3/a.

Sullivan (2014, p. 193) indicated the distribution of the OSRB water resources, as well as a comparison of areas and runoff in different parts of the basin (*Table 3.1*).

Table 3.1 Comparing area and runoff in different parts of the OSRB

River section or tributary	*% Of OSRB area*	*% Of total runoff*	*Runoff per area/ tributary (mn m^3/a)*
Vaal River basin	19	36	4,300
Senqu River to Lesotho South Africa border	2	34	4,010
Caledon River Welbedacht Dam	2	11	1,240
Orange upstream to confluence with the Vaal	6	11	1,300
Lower Orange excluding Fish River basin	64	4	420
Fish River basin	7	4	480
Total basin	100	100	11,750

Tributaries in the basin include the Caledon and the Vaal Rivers (in South Africa), the Fish and the more ephemeral Nossob and Molopo Rivers (in Botswana and Namibia); all of these are sizeable rivers. The Molopo forms the border between Botswana and South Africa; the Caledon forms the northern and western borders between South Africa and Lesotho; and the Orange River forms the border between Namibia and South Africa (*Figure 3.2*) (Sullivan, 2014, p. 190).

However, the Orange River's most important tributaries are the Senqu in Lesotho, the Vaal in South Africa, and the Fish in Namibia (Wirkus and Böge, 2006, p.23).

The basin is highly developed, with many dams and transfer schemes harnessing and controlling its flow. The OSRB is one of the most highly developed basins in all of southern Africa, with >300 built structures ranging from inter and intrabasin transfer schemes to large dams (*Southern Times*, 2015). There are 37 large-scale dams (height >25 m and storage capacity >12 million m^3) in the basin, most of them (24) are located in South Africa (Wirkus and Böge, 2006, p. 25).

However, being the most developed basin comes with a great deal of consequences. The high number of large-scale dams poses numerous social and environmental challenges. These challenges will be discussed in detail in the following sections of the chapter.

An analysis of environmental scarcity in the OSRB

Population growth affecting water resources in the OSRB

A key issue causing hydropolitical vulnerability in the Southern African region is population dynamics that have an impact on water resources (Earle, 2005, p. 60). Homer-Dixon (1999a, p. 112) argued that the challenge population dynamics pose to water resources is not only a regional issue but a global one as well.

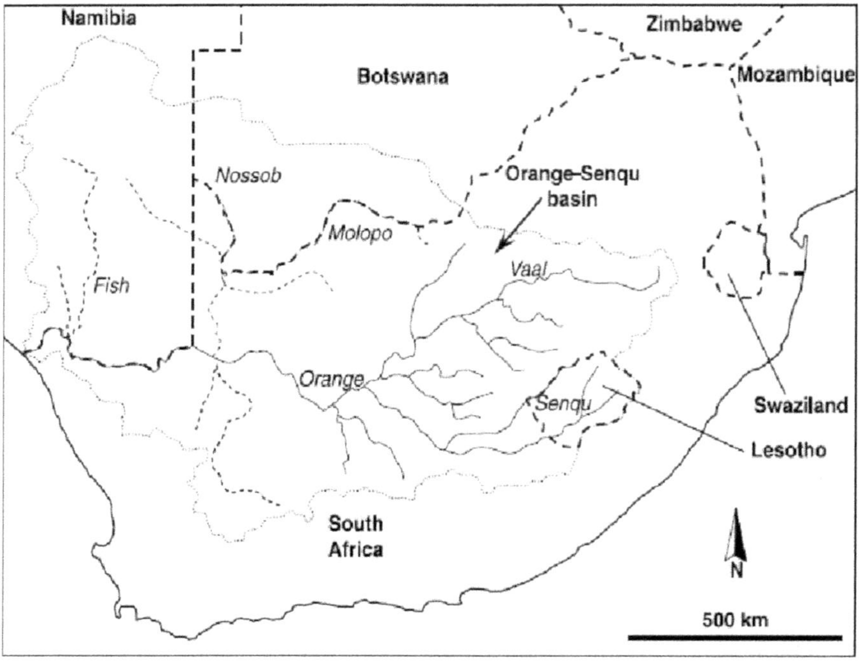

Figure 3.2 Map of the location and extent of the OSRB and the major tributaries in the basin.

According to the demand-induced scarcity argument, population growth leads to an increase in the consumption levels of water resources; therefore, decreasing the amount available to each individual, group, or state (Bingham, 2001). This is evident in the Southern African region. Rapid population growth in southern Africa is affecting the water needs of the region (Earle, 2005, p. 60). According to the latest United Nations (UN) estimates, the current population of southern Africa is 63 million (Worldometers, 2016). Growing industrialisation and urbanisation, as well as population growth, are placing further demands on water resources (Kranz *et al.*, 2005, p. 3).

The increased water demand in southern Africa is driven by high population growth (2%–3.5% per annum), urbanisation, improved living conditions, and industrial and agricultural development (Kranz *et al.*, 2005, p. 3). Increasing water demand and water use obviously result in less water being available per person. A theoretical overview of the link between environment, conflict, and population growth points out that the environmental effects of human activity are a function of two factors: the vulnerability of the ecosystem; and the product of the total population and that population's physical activity per capita in the region (Homer-Dixon, 1991, p. 85).

As asserted earlier in the book, Homer-Dixon's environmental scarcity theory argued for the existence of a strong relationship between population growth and environmental conflict. Homer-Dixon's environmental scarcity theory argues that population growth causes environmental scarcity (demand-induced scarcity). The conflict between various environmental users dependent on a specific resource facing depletion and degradation caused by population growth may ensue as a result of such scarcity (Homer-Dixon, 1999b, p. 61). The theory emphasises the notion that population growth, density, or both are often cited as significant conflict risk factors (Homer-Dixon, 1999a, p. 48).

More than 19 million people rely on the OSRB (*Southern Times*, 2015; INR, 2016). In line with Homer-Dixon's environmental scarcity theory, this chapter argues that high population growth and high dependence on water resources can decrease the supply of water by calling for the increased construction of large dam infrastructure to meet demands and contributing to resource depletion and degradation through unsustainable use (Homer-Dixon, 1999a, p. 52).

Certain areas of the basin are densely populated (Orange-Senqu River Commission (ORASECOM), 2008, p. 2). The proportion of each country's national population that is located within the Orange Senqu Basin and relies on its water supplies varies greatly among the states (Kistin, 2010, p. 139). The Vaal catchment is highly populated and urbanised. Similarly, most of South Africa's heavy industry and mining activities are situated within this catchment (Jacobs, 2009, p. 98).

The DWS stipulated that the demographics of the Vaal River catchment have changed significantly over the years, with increasing numbers concentrated around the urban areas. As a result, the estimated population for the Vaal area[1] is close to

11 million, but the Vaal River system supplies water to a total population of about 20 million (South African Department of Water and Sanitation, 2014:6).

Moreover, virtually the entire population of Lesotho, estimated at 2.1 million (*Country Meters*, 2016), is resident in the basin (Jacobs, 2009, p. 98). Lesotho benefits immensely from significant water resources, but ironically, in the capital Maseru, water stress is becoming an enduring reality largely due to population growth. This population growth is a result of migration from the highlands to the lowlands (Maseru) largely due to the industries found in the capital. This migration translates into an increased water demand. Subsequently, this requires various forms of water transfer, one of which is the Metolong Dam. While the water resources originate in the highlands, agricultural opportunities and industrial growth is found in the lowlands (Sullivan, 2014, p. 194–195).

It is conceivable to argue that this growth in population and the density of the population in the OSRB will affect resource availability. Theoretically, this growing population and greater per capita resource demand can lead to increased demand, supply, and structural-induced scarcities (Homer-Dixon 1999a, p. 52).

When the dependence of a large population on a shared resource increases, that resource will face depletion (because of overuse) and degradation (because of pollution caused by unsustainable use). Moreover, powerful water users will eventually attempt to capture large portions of the resource (through institutions, agreements, plain force, or both) to meet their growing demand; therefore, structurally denying access to other water users (Homer-Dixon, 1999a, p. 52).

Accordingly, and in line with the arguments of Homer-Dixon's environmental scarcity theory, this increase in population growth will result in two phenomena essential to conflict. First, this growing population will lead to an increase in demand for a resource, which then assumes that the constant use of this resource will result in its depletion (Homer-Dixon, 1999a, p. 48).

Since water resources are rivalrous (when the use of a resource by one group reduces its availability for others), this will result in competition between different water users and lead to disputes and conflict (Percival and Homer-Dixon, 1998, p. 280). Similarly, Van Schaik and Dinnissen (2014, p. 33) contended,

> The risk for inter-state conflict increases when the reinforcing nature of population growth and environmental stresses coincide with the weakening of regimes.

Degradation and depletion of water resources in the OSRB

Water availability in the OSRB is already at a critical stage. When combined with the relatively high water demands and withdrawals for agricultural use, this river basin is among the most severely water-scarce river basins in Africa (Diederichs *et al.*, 2005, p. 24). Across the entire basin, water stress is widespread (Sullivan, 2014, p. 194). Future projections of the various physiographic, socioeconomic, and management data suggest further aggravation of the situation. Being the most developed river in Africa, the Orange River is at present nearly fully exploited for a variety of purposes, and South Africa is the country most dependent on its water (Diederichs *et al.*, 2005, p. 24).

The OSRB riparian countries rely to varying degrees on the river as a source of water for industry (e.g., mining and manufacturing), agriculture, and domestic uses. Consequently, the basin is degraded environmentally and continues to be threatened as a result of industrial, agricultural, and domestic factors (ORASECOM, 2008, p. 2). This situation may lay the foundation for conflict in the absence of effective regimes (e.g., properly functional governance systems, such as institutions meant to protect and preserve shared natural resources) coupled with poor political will (Young, 2011, p. 19854).

Increasing population growth, urbanisation, industrialisation, a lack of maintenance of the supply infrastructure, such as dams and pipelines, and the anticipated effects of climate change on river flows are some of the geophysical and socioeconomic factors further constraining the OSRB, which are already completely utilised. Mwangi (2008) argued that such human-induced environmental degradation threatens human and environmental security. Similarly, Homer-Dixon's environmental scarcity theory argues that large-scale human-induced environmental pressure may seriously affect national and international security (Homer-Dixon, 1991, p. 76).

The depletion and degradation of resources come as a result of the physical vulnerability of the resource, the size of the population consuming the resource, and the consumption behaviour (Homer-Dixon, 1996, p. 360). This is true for the OSRB. The OSRB has been depleted and degraded because of its vulnerability to natural and human-induced environmental factors, such as climate change, industries, agriculture, and domestic use. It is also confronted with a rapidly growing population that is largely dependent on its resources and faces negative consumptive behaviour because of large dam infrastructure across the basin and inter- and intra-basin water transfers IBTs.

The World Resources Institute's Aqueduct online identified the OSRB as a hot spot for water risk. Droughts and insufficient rainfall contribute to what's known as water risk, along with floods and contamination (Zabarenko, 2011). Most importantly, riparian states in the OSRB are under high levels of water stress. Increasingly, water scarcity limits these states' ability to meet future demands (Sebastian, 2008, p. 147), making tensions and disputes over water resources probable.

Water availability is a main transboundary issue in the region (Jacobs, 2009, p. 98). Water scarcity poses a serious challenge to the basin in view of the climate of the entire basin (Mokorosi, 2007, p. 3). Such scarcity has also placed limits on the future economic growth potential of the region (Earle, 2005, p. 56). Each of the riparian states faces water constraints and relies increasingly on shared international water resources (Langea *et al.*, 2007, p. 660). Of the four riparian states of the Orange Senqu River, three are classified among the driest countries in the Southern African Development Community, for example, South Africa, Botswana, and Namibia (Jacobs, 2009, pp. 98–99).

It is estimated that in 2020 water demand will outstrip supply by 93% in the region and that by 2025 South Africa will face water scarcity,[2] Lesotho will be water stressed,[3] while Botswana is likely to experience water quality and availability problems in the dry season. Although Namibia, the most arid country in the basin, is considered to have adequate supplies of water due to its small population, significant variations exist in the supply to populations (Diederichs *et al.*, 2005, p. 43).

Despite these estimates, Sullivan (2014, p. 195) noted that Lesotho is not expected to experience either water stress, scarcity, or both in the near future given its abundance of water, and its current small water consumption. This means that there is still room to further develop water use for income generation, including electricity (Sullivan, 2014, p. 195). Nonetheless, in this context of shared river basins it can be argued that when the water demand outstrips the available supply, as estimated by Diederichs *et al.* (2005, p. 43), a nation will be able to justify military action (whether offensive or defensive) in the name of economic preservation and national security (Van Schaik and Dinnissen, 2014, p. 33).

According to ORASECOM (2008, p. 129),

> the OSRB is one of the most modified and degraded in southern Africa.

This is a result of inappropriate management, which has contributed to numerous issues such as:

- the reduced performance of the catchment in terms of a reduction in water yield.
- deteriorating water quality.
- increased sediment loads because of erosion due to poor rangeland management and cropping practices.
- a loss of biodiversity.
- lowered land productivity.

All of these have led to socioeconomic hardships (ORASECOM, 2008, p. 129).

Water quality in the basin is impaired by seepage, runoff, and point source discharges of municipal, industrial, and agricultural effluents, and by high sediment loads resulting from land degradation in many areas of the catchment (ORASECOM, 2008, p. 2). Moreover, the sedimentation problem in the OSRB is caused by natural factors, such as easily eroded sandstone and rainfall but aggravated by human factors, such as a growing population. The sedimentation problem is the worst in South Africa (Burnett, 2009).

Damage to wetlands high in Lesotho's Maluti Mountains posed serious challenges to the quality and quantity of the entire Orange Senqu River system. With wetlands crucial for retaining, purifying, and regulating the flow of water in this high rainfall area, the degradation of the wetlands affects the water supply of the Lesotho Highlands Water Project (LHWP). The LHWP captures water in dams and supplies it to the water-thirsty South African industries and agriculture (Burnett, 2009). However, a combination of factors, including infrastructure development, overgrazing, and cultivation and the resulting erosion, led to the wetlands being degraded (Burnett, 2009). In addition, the erosion contributed to increased sedimentation of water downstream of the wetlands and, thus, in the Orange Senqu River system (Burnett, 2009).

Droughts are an important issue for all countries within the basin. South Africa, as the country most dependent on the water of the Orange River, is especially affected, but has some reaction capacity because of extensive damming and the existence of water transfer infrastructure, within South African river basins as well as with Lesotho. Despite the relatively high rainfall, droughts and desertification are also an issue for Lesotho, especially in the southern districts of the country. Botswana and Namibia are, due to their water resources stress, clearly very vulnerable to droughts. However, Namibia, due to its downstream riparian status, is especially vulnerable (Kranz *et al.*, 2005, p. 3). With the current El Niño, Nina events, or both it is evident that these droughts are expected to persist for some time.

Since Lesotho lies entirely in the basin, the negative human and natural impact on the environment is cause for concern for the entire basin. According to Water Affairs Minister Kimetso Mathaba,

> Massive land degradation in catchment areas is increasingly threatening the sustainability of water, land and related resources. ... Although the country was blessed with abundant water resources, these were increasingly threatened by climate change which has been manifested among other things, through the prolonged and recurring El Nino-induced drought
>
> (Lesotho Times, 2017)

Rangeland degradation (i.e., loss of vegetation cover) is also evident in the OSRB. This ranges across the loss of grass species diversity and perennial grasses, a loss of grass vigour and ground cover, decreased soil moisture, soil crust formation, and declining land productivity leading to increasing vulnerability to drought. In the south of Namibia, Lesotho, and parts of South Africa, rangeland degradation is by far the dominant expression of degradation.

There are generally two interrelated causes for the loss of vegetation cover. The first cause is overstocking, which describes a situation where more animals are kept on a certain piece of land than there is fodder available to feed them. The second cause is overgrazing, which is caused when animals are concentrated in one specific area for too long, resulting in the vegetation being overused without adequate recovery time. Open access to land, unsuitable distribution of water and boreholes, and a lack of institutional mechanisms and management skills are all major factors in overstocking and overgrazing (ORASECOM, 2008, p. 129).

Deforestation is the second major form of land degradation and refers to the removal of woody vegetation cover. One of its consequences is a change in the hydrological cycle. Deforestation is prevalent in Lesotho and along the riparian belt of the middle and lower Orange River, because of the limited woody resources in the more arid areas and clearing for agricultural purposes. The predominant fuel for cooking, heating, and lighting in communal areas and low-income urban areas remains wood fuel (ORASECOM, 2008, p. 129).

Due to the high level of land degradation in the OSRB, potential links between land degradation and conflict seem likely. However, as noted by Van Schaick and Dinnissen (2014, pp. 11–12), one should not think of a direct link, relationship, or both but rather of how land degradation is related to conflict risk factors. While the relationship between land degradation and conflict exists, it is a complex one and is indirect rather than direct.

The OSRB is characterised by the development of industries such as mining, and hydro infrastructure construction, such as dams and IBTs. Mining (e.g., extraction and processing) is recognised as negatively impacting the environment, particularly on water resources (Meyer, 2013, pp. 31–32). Sullivan (2014, p. 194) posits that in the Vaal tributary, mining and industry provide the biggest pressures, compounded by the inevitable pollution from growing urban and peri-urban populations. Similarly, in recent years, Lesotho has witnessed developments in the mining sector, such as the Letšeng and Liqhobong diamond mines. Although these developments have been good for the country's economy, this development and the potential pollution associated with it will put additional stress on water resources at the local level (Sullivan, 2014, p. 195).

The most frequently mentioned impact of mining is acid mine drainage, which arises when the mineral pyrite meets oxygenated water. As the rock mass becomes extensively fragmented during mining, the surface area is greatly increased, as is the rate of acid production. The truly detrimental impact of mining arises when large quantities of acidic water are released into the environment, polluting surface and groundwater resources with heavy metals. Lately, the Vaal catchment is considered to be highly polluted, affecting water resource availability for other economic sectors within the basin and the population (Meyer, 2013, pp. 31–32).

The construction of large dams in the basin in response to the growing water demand from agriculture, industry, and domestic water use further aggravates the quality and quantity of the basin. The construction of an extremely intricate network of IBTs is causing further complications and hydropolitical vulnerability in the basin (Earle, 2005, p. 60). It is sufficient to argue that countries currently developing water resources through hydro infrastructure construction, such as dams and IBTs are doing so in response to pressures from population growth and economic development (Veilleux, 2013, p. 1). However, the challenge is that the social and environmental implications on inhabitants associated with these projects are far greater than the hydro power potential expected. For instance, in Lesotho, the construction of huge dams took over people's valleys, homes, fields, gravestones, and grazing lands, all of which were ultimately submerged by water (Bennett, 2012).

First, the challenges posed by the construction of large dams include uneven water distribution for other riparian countries as a result of over-extraction (Diederichs *et al.*, 2005, p. 24). Second, they include massive displacements. Displacement is commonly associated with refugees fleeing conflict; however, millions of people are resettled every year because of *development and progress*. Many of these individuals endure social and cultural disruptions as well as economic upheaval, and a significant number never regain their former quality of life. In the 1990s, there was a growing recognition that the vast majority of those displaced by large development projects failed to retain their standard of living. The reality was that most slid into greater impoverishment (Bennett, 2012).

Third, these projects result in the loss of arable and grazing land and households. For instance, in Lesotho alone, a country with an environment characterised by a number of problems, such as steep slopes and frail soil formation, the decline in soil quality, soil erosion, the scarcity of arable land, and climate variability were also affected by these loses. Lesotho lost approximately 2,700 ha of grazing land and 925 ha of arable land, with 71 households being displaced and 20,000 people being affected because of the construction of the Katse Dam in 1995. Moreover, following the completion of the Mohale Dam in 2007, 1,125 ha of grazing land and 875 ha of arable land were lost, with 325 households displaced in stages 1 and

2 of the project. In total, 71 households also lost>50% of their land, and 7,400 people were affected (Henwood and Funke, 2002, p. 182; Furlong, 2006, p. 452; Mwangi, 2007, p. 15; Mwangi, 2008; Hitchcock, 2015, p. 528).

Due to increasing water scarcity and the predicted levels of water stress to confront certain riparian states, there will be further calls for the construction of dams and ever-greater water transfer schemes. Although this can bring a measure of temporary relief (Earle, 2005, pp. 60–61), the negative socioeconomic impact of these projects is greater.

With 90% of the LHWP infrastructure located in Lesotho, the environmental and, specifically, the social impact of the project were incurred by that country (Mokorosi, 2007, p. 9). As indicated previously, the LHWP resulted in the destruction of arable land and posed a threat to local food security and life dependent on the natural environment, a process termed *ecocide* (Mwangi, 2007, p. 11). As defined by Lee (2001, p. 75), ecocide is the deliberate destruction of the natural environment through various human activities that endanger human life. Ecocide may be the result of externalities, such as pollution, which destroy the ecosystem, or corporations and governments operating on indigenous lands using less-than-adequate safety procedures (Lee, 2001, p. 75).

According to Muller *et al.* (2015, p. 103), the threats to regional integration are that the failure to agree to measures to address scarcity or the impact of upstream activities such as pollution could promote interstate conflict. Lee (2001, p.75) further argued that pollution leads to a situation where the lands, reproductive ability, and long-term health of the indigenous population are irreparably damaged. Ecocide also results in displaced communities, sometimes referred to as environmental refugees, and environmentally-induced conflicts (Lee, 2001, p. 75).

According to the Transformation Resource Centre (TRC) (2003, p. 3), the destruction of arable land has caused the displacement of large numbers of people (environmental refugees) without compensation from the Government of Lesotho. This creates a long line of poverty, which results in a threat to the economic security of Lesotho. The threat to security has occurred despite the existence of the LHWP Treaty, specifically Article 15 which deals with the regard to be given to social and environmental considerations by the Lesotho Highlands Development Authority (LHDA), which is the implementing agency of the LHWP. Article 15 states,

> The Parties agree to take all reasonable measures to ensure that the implementation, operation and maintenance of the Project are compatible with the protection of the existing quality of the environment and, in particular, shall pay due regard to the maintenance of the welfare of persons and communities immediately affected by the Project.
>
> (Kingdom of Lesotho-RSA, 1986, p. 10)

Exacerbating the issue is the fact that Namibia is also interested in building a new dam on the Lower Orange River (Heyns, 2004, p. 7).

Most of the OSRB water is used for irrigated agriculture. Agriculture alone accounts for ≤80% of the total water use in the OSRB (Sullivan, 2014, p. 193). In South Africa, for instance, agriculture dominates much of the main branch of the Orange River (Sullivan, 2014, p. 194). The high dependence on local natural resources, such as agriculture makes the economies and social character of the OSRB particularly vulnerable to changes in the availability of water over space or time. Such changes may be defined in terms of the total amount of precipitation received its frequency of recurrence, the persistence of wet or dry day combinations, the onset and duration of the rainy season, or in terms of the quality of the available resource. The extent to which water resources, the environment and economies may be affected by changes in water availability varies.

In South Africa, 800,000 ha of land is presently under irrigation: and 300,000 ha of this total is located in the Orange River Basin (Heyns, 2003, p. 21; Heyns, 2004, p. 8). However, some urban centres and mining projects are also among the region's major water consumers. The problem of water quantity is increasingly aggravated by a water quality problem caused by pollution from towns, industry, mining projects, and saline return flows from irrigation projects.

Namibia is the most downstream riparian state with a high reliance on the Orange Senqu for agricultural activity in the south of the country, even though a relatively small proportion of its population (8.9%) live in the basin's territory. Namibia has an extremely arid hydroclimate, a high level of water stress, and is unique in that all its perennial rivers are transboundary. These characteristics make it particularly vulnerable to external dynamics regarding the river and Namibia, therefore, relies heavily on international water resources to meet internal demand. Specifically, Namibia relies on South Africa for future water storage developments to increase its assurance of supply. In the southern parts of Namibia, the greatest development potential lies in irrigation, which subsequently creates the highest demand for water (Kranz *et al.*, 2005, p. 3).

For instance, Namibia has plans to sharply expand irrigated agriculture on the Lower Orange River (currently some 2,200 ha) to produce grapes for the European and North American markets.

> It is clear that the future development of the labour intensive export grape industry is economically viable, but is heavily dependent on the availability of water.
>
> (Heyns, 2004, p. 7)

Future irrigation developments will hinge in crucial ways on an adequate, secure supply of water. In this case, the use of *drip (trickle) irrigation systems* may be beneficial.[4]

However, it is not certain that farmers will accept the new method. Servicing the equipment may be problematic and the costs may be high compared to the benefits. Moreover, drip irrigation is sometimes more expensive. This is influenced by various factors that affect the cost such as land size and quality of parts. Similarly, the construction, installation, operation and maintenance (per ha) should also be taken into account (Brouwer *et al*, 2001:59).

> construction, installation, operation and maintenance (per ha) should also be taken into account
>
> (Brouwer et al., 2001, p. 59)

As per the Intergovernmental Panel on Climate Change findings, Swain and Krampe (2011, p. 17) indicated,

> Southern Africa will become much hotter (minimum plus 4°C) and drier (about 10-20% less rain).

Climate change is also one of many security threats facing the OSRB. The impact of climate change is likely to produce a host of social and political problems that are likely to weaken OSRB riparian member states and societies, thus laying the foundation for conflict. This is due to the existing water scarcity, which is further aggravated by the impact of climate change. While the OSRB is characterised by high geographic variability, large agricultural and industrial demands, and historic inequities in supply, the basin is also characterised by natural climatic variability, which is normally characterised by naturally variable rainfall patterns with frequent periods of floods and drought. This is a key issue causing hydropolitical vulnerability (Earle, 2005, p. 60).

Temperature increases and seasonal shifts in the OSRB have had various impacts on the environment and the population. The consequences of seasonal shifts included more floods during the rainy season and longer dry periods, which affects the water supply of the basin (Meyer, 2013, pp. 52–53).

Future predicted effects of climate change will include water scarcity, which affects people's security; impacts on agriculture due to shifts in seasons, which affects food security; a rise in natural disasters such as field fires, droughts and floods, changes in water runoff, and precipitation due to snow melt in the Lesotho highlands; and higher water pollution levels during the dry season (Meyer, 2013, p. 53).

Climate variability in the basin is seen as a natural phenomenon, which varies in space and time. Although to a certain extent people are already used to some climate variability and the consequent impact on livelihoods, climate change is becoming more and more problematic (Meyer, 2013, p. 55).

Declining aggregate rainfall could have a significant impact on agricultural practices, trade, and the average quality of life in the basin. In a basin where water supplies are decreasing and where use will likely soon outstrip supply, the effects of global warming could have profound and potentially devastating effects across a range of uses, particularly in the agricultural sector. Observed changes in temperature and precipitation are very likely to change the land cover and affect land use patterns in the basin, resulting in shifts in the supply and quality of ecosystem services due to altered flow regimes and crop production, biodiversity loss, and the introduction of increased alien species (Diederichs *et al.*, 2005, p. 24).

The hydrological system is likely to see some of the most significant impacts of global climate change (Diederichs *et al.*, 2005, p. 72). Some of the adverse impacts of climate change, in combination with a number of political and socioeconomic factors, are likely to contribute to an increasing potential for conflict (German Environment Ministry, 2002, p. 4). Southern Africa, including much of the OSRB, is prone to drought and flooding, and its water resources are highly sensitive to changes in rainfall and other climate variables (Diederichs *et al.*, 2005, p. 72). However, these droughts and floods may not directly cause conflict but may trigger or aggravate it by overburdening states and regimes (Tadesse, 2010, p. 8; Swain and Krampe, 2011, p. 17).

According to the CIA World Factbook (n.d.), the existing disagreements surrounding water allocation in the Lower Orange River Basin might be further exacerbated by the impact of climate change. The basin has long experienced substantial seasonal and spatial variation in rainfall and evaporation rates. While annual precipitation at the source of the river (i.e., in the Lesotho highlands) is relatively high and evaporation rates are low, the opposite is true for the downstream portions of the basin. Forecasts suggest that the existing spatial variability will be further increased by the impact of climatic change. Especially in downstream areas of the basin, precipitation is expected to decrease, and the resulting diminished surface runoff and groundwater recharge are likely to pose serious challenges when maintaining food security and economic development in South Africa and Namibia.

Uneven distribution of water resources in the OSRB

Water availability and allocation are probably the main transboundary issues in the region (Earle *et al.*, 2005, p. 13). According to Meyer (2013, p. 30), the scarcity and unequal distribution of freshwater resources in the OSRB are considered one of the fundamental threats to the economic and social development in the Southern African region. According to 1Sullivan (2014, pp. 93–194), South Africa withdraws 63% of the freshwater that is available in the basin. Namibia's use amounts to only 1.3% of the available water in the basin. Lesotho, despite its plentiful supplies, only uses 0.2% of the OSRB's water.

It is important to note that the construction of large dams and IBTs in the region, in response to the growing demand as a result of population growth, has led to further uneven distribution of water, water diversions, or both (Earle, 2005, p. 60). The fact that the urban and industrial demands are geographically concentrated in the upper parts of the basin (i.e., South Africa) and that these demands support activities that make a major contribution to the gross domestic product (GDP) of South Africa (the largest basin state) creates a significant geographical imbalance in the use of the available water resources (ORASECOM, 2008, pp. 2–3).

South Africa is highly economically dependent on the Orange Senqu River for a staggering 100% of the GDP of Gauteng, which is the industrial and economic heartland of South Africa and arguably Africa. As a result, South Africa is dependent on IBTs involving the Orange system (Jacobs, 2009, pp. 98–99); therefore, consuming approximately 60% more water than other riparian countries, as witnessed earlier in the chapter. Moreover, South Africa may be driven to pursue extensive resource capture (mostly using the LHWP Treaty), considering the future water scarcity estimates of South Africa.

For this reason, the Vaal has been linked via a complex system of IBTs with eight other river basins, tapping them to meet Gauteng's water needs. This has meant that the Orange–Vaal system has a negative water balance. At the same time, economic activities in Gauteng are 100% dependent on IBTs. This, in turn, means that the Orange River has huge strategic significance for South Africa, a fact which casts some light on the importance of the LHWP (Turton *et al.*, 2004, p. 99).

Lesotho exports water to South Africa at an LHWP Treaty's stipulated rate of 70 m^3/s, which literally translates to 4.2 million L of water/min (News24, 2016). Due to this water export, Lesotho has become highly dependent on South Africa's royalties for IBTs for its economic development. With the lowest GDP of all four riparian states, the export of water forms the majority of Lesotho's export revenue (Jacobs, 2009, p. 98). However, it is essential to note that regimes, such as the LHWP Treaty can, and have, become sources of structural-induced scarcity. This notion is highly supported by Homer-Dixon (1999a, p. 15) who argued that the imbalance in water distribution is often and deeply rooted in regimes, institutions, or both.

Although the LHWP is a notable example of interstate political and economic cooperation, it also poses a possible source of future interstate conflict. Despite Lesotho being a water-abundant country, scenarios indicate that the country faces water stress and scarcity because of climate change. This creates the potential for future interstate conflict as Lesotho may be unable to supply adequate water to South Africa under the prescribed terms outlined in the LHWP Treaty, as Lesotho will have a pragmatic and moral obligation to first meet its own domestic needs. However, there is a strange aspect to the LHWP Treaty.

Paragraph 8, Article 6 of the LHWP Treaty states,

> Neither of the parties shall cause or permit under any circumstance nor for any reason whatsoever any unilateral interference with the delivery of water to the Designated Outlet Point.

While all this water is being delivered from the LHWP into the Vaal Dam, >70% of all taps in the kingdom are dry simply due to the lack of good rains that, in most instances, last up to about 12 months as a result of El Niño (News24, 2016).

The unfortunate implication of this is that regardless of how long the so-called El Niño may last, and regardless of how often and how much worse it gets, there is no way in which Lesotho can momentarily reduce, let alone stop, the outflow of its water into the Vaal Dam solely to ensure that domestic demand is met first. This situation has

> given rise to disputes and tensions. Water conflicts between the two riparian states have, arguably, already occurred, despite the current abundance of Lesotho's water
>
> (News24, 2016)

Therefore, arguing based on the structural-induced scarcity perspective, Lesotho will experience water scarcity, not in the sense of actual scarcity, but the sense of a lack of full use of the basin. This is simply because the LHWP Treaty put in place to regulate water sharing is perpetuating uneven water distribution. According to Homer-Dixon (1999a, p. 15), resource scarcity is a result of a *severe imbalance in distribution* and is a key factor in virtually every case of scarcity contributing to conflict. They further asserted that as in the case of Lesotho–South Africa water relations, imbalance in distribution is deeply rooted in institutions

Kreamer's (2012, p. 90) argument that *perceived misallocation of resources has instigated clashes between water users* is also true in the case of Lesotho and South Africa, with reference to the 1986 water coup and the 1998 South Africa-led Southern African Development Community (SADC) intervention in Lesotho. These events may repeat themselves, although not under the same circumstances if Lesotho attempts to reduce the flow of water to South Africa, which would be a gross violation of the LHWP Treaty. The government of South Africa may find itself with no option but to respond (News24, 2016).

According to Homer-Dixon (1994, p. 19), on 20 January 1986, South Africa gave decisive assistance to a successful military coup that toppled Lesotho's government. South Africa had numerous reasons for ousting Prime Minister Leabua Jonathan, including the fact that Lesotho had offered sanctuary to African National Congress (ANC) guerrillas. However, one key motivation is often overlooked: South Africa sought greater access to Lesotho's water supply.

The South African province of Transvaal faced critical water shortages, and, despite 30 years of negotiations, the South African government could not reach an agreement with Lesotho for water rights. Within months of the coup, both governments agreed to the Highlands Water Project, which diverts water from Lesotho's mountainous regions to South African farms and industries. The timing of the agreement suggested a close link between South Africa's involvement in the coup and the dispute over access to water (Homer-Dixon, 1994, p. 19).

Furthermore, as demonstrated by the 1998 South African-led SADC intervention in Lesotho (also referred to as *Operation Boleas* due to the nature and character of the intervention (Neethling, 1999, p. 1) following political instability as a result of alleged election fraud, it is evident that South Africa's vested interest in the intervention was water (Bischoff, 2003, p. 196). Although the political instability in Lesotho occurred in Maseru, the South African National Defence Force special troops went to the Katse Dam (211 km from Maseru) and killed 18 Lesotho Defence Force members guarding the dam (Likoti, 2006, p. 166).

Moreover, the OSRB is characterised by historic inequities in supply because of border demarcation, water allocation and pricing, subsequently leading to disputes (CIA World Factbook, n.d). According to Kranz *et al.* (2005, p. 3),

The indisputable facts are that South Africa has the most control over the Orange River and that Namibia is the hardest hit.

As a result, this has constantly led Namibia to question the border arrangement between itself and South Africa; therefore, intensifying the border dispute. According to the CIA World Factbook (n.d), in addition to the common border demarcation that Namibia is contesting, water allocation, and pricing, the LHWP has further aggravated the tensions between South Africa and Namibia.

The Orange River forms the border between South Africa and Namibia. There are some problems concerning the actual course of the border between both countries (Turton *et al.*, 2004, pp. 99–100). In 1890, the boundary was defined based on the high water level along the northern banks of the river, and this meant that Namibia was without direct access to the river's waters (Turton *et al.*, 2004, pp. 99–100).

On the one hand, in terms of the Namibian constitution, Namibia claims,

> The national territory of Namibia shall consist of the whole of the territory recognised by the international community through the organs of the United Nations as Namibia, including the enclave, harbour and port of Walvis Bay, as well as the off-shore islands of Namibia, and its southern boundary shall extend to the middle of the Orange River.
>
> (Poolman, 2014)

On the other hand, South Africa claims, based on the 1890 treaty, that the border runs along the north bank of the Orange River, and its Recognition of the Independence of Namibia Act denies any recognition of Namibia's claim.

Before independence, Namibia was promised that the boundary would be shifted to the middle of the river; however, the promise was not respected and, since then, there have been disputes over the border (Turton *et al.*, 2004, pp. 99–100). According to the Southwest Africa People's Organisation's (SWAPO) Secretary-General Nangolo Mbumba,

> Officials of the Surveyor-General of both countries were involved with the demarcation and the then (apartheid) South African regime agreed that the border between the two countries run through the middle of the river. Both countries have minutes to this affect. It was assumed that this initial arrangement on the Orange River was finalized, but unfortunately no agreement was signed at that stage.

However, the South African ANC government has a different view on the river, arguing that its northern bank constitutes the border between both countries.

This would mean that Namibia would not have any access to the river or its water (Poolman, 2014).

The assertion by the ANC government and the position of Namibia on this issue could lead to conflict. This is because, according to the structural-induced scarcity perspective, as water resources are *excludable*, meaning that property rights or other institutions can be used to prevent access to the resource by some actors, they are likely to lead to conflict. Renewable resources, such as river water have characteristics that permit the assignment of clear property rights. Conflicts over natural resources arise when parties disagree about the ownership, allocation, and use of natural resources (Homer-Dixon, 1999a, p. 48).

This is despite a gentlemen's agreement, following South Africa's independence in 1994, reached between the then presidents of Namibia and South Africa, Sam Nujoma and Nelson Mandela, respectively. This agreement stipulated that the people living on both sides of the border would have access to the water resources in accordance with international law and the 1995 SADC protocol on shared watercourses, which requires that communities of riparian states have access to common water sources (Poolman, 2014).

Namibian Foreign Affairs Minister Netumbo Nandi-Ndaitwah confirmed that the dispute is still dragging on, irrespective of President Hifikepunye Pohamba discussing the Orange River dispute with his South African counterpart Jacob Zuma during a state visit to Namibia early in 2014. A working group between both countries has been established to find an amicable solution. Nonetheless, Namibia's Foreign Minister does not seem to be very optimistic that this thorny issue would be solved soon (Poolman, 2014).

Furthermore, at the time of Namibian independence in 1990, it was agreed that Namibia would receive a certain amount of water from existing infrastructure in South Africa to meet its water demands downstream. Namibia would receive this share of water, which it regards as its historical allocation, free of charge. Over the years; however, Namibia has sought to develop its southern region, and the only possibility to promote socioeconomic development in the south would be by using a greater share of the Orange River waters. Namibia has additional water needs for irrigation (i.e., table grapes for export), mining (e.g., zinc and diamonds), and power generation (i.e., development of the Kudu gas field). To meet these water requirements, Namibia has asked for an additional temporal allocation of water from South Africa (CIA World Factbook, n.d).

South Africa has responded to this demand by suggesting that any additional water allocations to Namibia should be charged at full cost according to South African water tariffs. In addition, South Africa has made clear that the historical allocation that was taken for granted by Namibia should be subjected to operation and maintenance costs incurred by South Africa for water storage and regulating infrastructure on the Lower Orange River (Shilomboleni, 2006, p. 11). Currently, Namibia does not have its water infrastructure on the Orange River and, in the past, has benefited from the South African infrastructure without sharing the burden of its costs. In the future, a dam tailored to the needs of

Namibia will likely be built on the lower reaches of the Orange River (Heyns *et al.*, 2008, p. 378).

However, further complicating the situation is the expansion of the LHWP, which Namibia fears will reduce water availability in its southern region. In light of these concerns, the Namibian government has made a request to become more involved in Phase II of the project. However, South Africa and Lesotho seem to prefer the existing bilateral mechanisms for communication and information sharing on the LHWP. The primary reasons given for their reluctance to open the LHWP up for greater Namibian participation include the efficiency of bilateral partnerships for project implementation, the complexities of including additional states, and the belief that existing mechanisms in the basin, especially ORASECOM, provide sufficient possibilities for Namibian involvement (Kistin and Ashton, 2008, p.397).

Water regime formation and capacity in the OSRB

According to Mwendera and Atyosi (2018, p. 266), water resources are not only becoming increasingly limited as economies and populations grow, but rainfall is highly variable and unevenly spread across the region. Frequent droughts, often broken by severe flooding, have affected lives, slowed economic growth, and posed significant problems for water managers. The intensive use of limited water resources has affected their quality and had an impact on downstream river ecosystems. Climate change and variability may further exacerbate these problems. Therefore, there is a need for increasingly close cooperation among nations that share river courses to continue to grow our economies, address poverty and protect the environment (Mwendera and Atyosi, 2018, p. 266).

Water scarcity in transboundary basins can constrain the economic and social development of the dynamic and growing economy in basins; therefore, leading to disputes and tensions that make adequate water resource management necessary.

Institutional capacity, be it in the form of river basin commissions and regional water structures, is defined as a legal framework in the form of formal treaties and protocols, technical cooperation in the form of informal working groups and technical task teams, or generally warm relations, has been prioritised as the heart of conflict management, particularly in arid countries (Wolf, 2005, p. 13).

An obvious way for states to cooperate over the management of shared waters is by negotiating international agreements (Hiddema and Erasmus, 2007, p. 2), which does not encroach on the sovereignty of states and allows them to reconcile national legal and jurisdictional aspects with regional legal and jurisdictional infrastructure (Kistin, 2012, p. 42). As regime theory itself argues in favour of respect for state sovereignty in matters relating to shared water resources, this chapter can assume that regimes are indeed essential tools for reducing uncertainty and mitigating the fear of exploitation among riparian countries (Hasenclever *et al.*, 2000, p. 7).

River basin organisations are at the core of the IWRM, with the mandate of building consensus and harmonising local, regional, and national water policies.

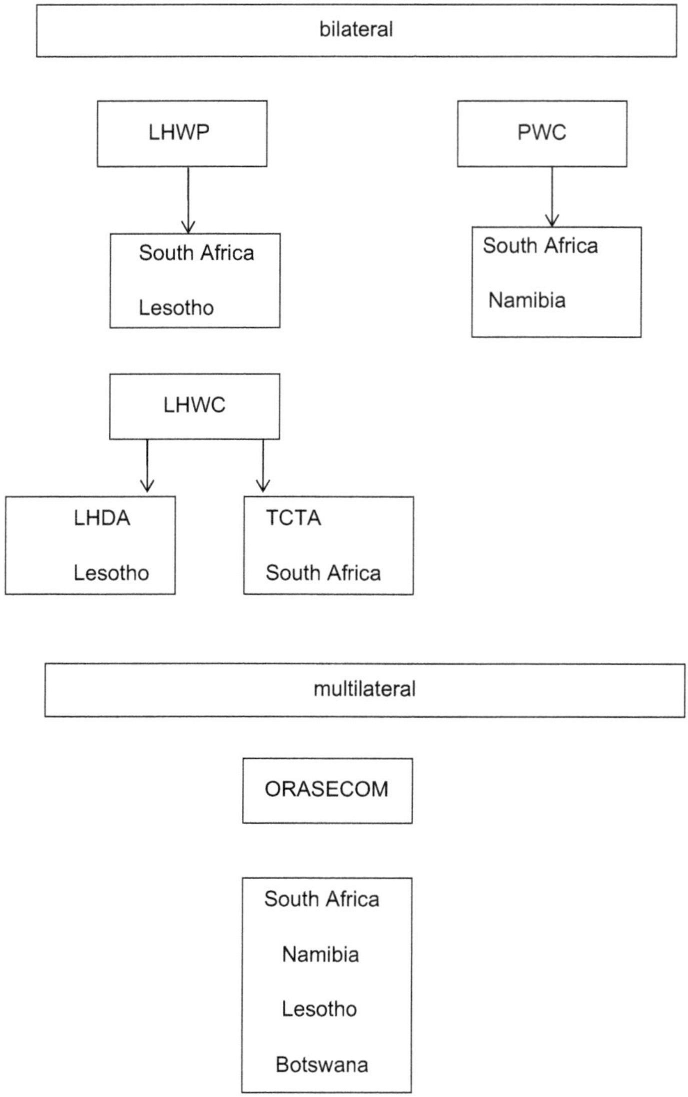

Figure 3.3 Overview of the Orange Senqu Basin governing institutions.

Effective water governance relies on national laws and institutions at the national, provincial, and local levels.

The OSRB is characterised by a rich history of interstate interaction over water resources and a multiplicity of international water management institutions that have emerged and evolved (*Figure 3.3*) (Kistin, 2012, p. 42). According to Jacobs (2009, pp. 100–101), negotiations over the waters of the OSRB have been ongoing between various combinations of the riparian states since the 1950s, which has resulted in a wide range of bilateral and multilateral interstate and government-led commissions and agreements, project-based organisations, treaties, and technical committees.

This cooperative framework has, to a large extent, been determined by the hydropolitical history of riparian relations and the domestic context of national regime types. In addition, an international policy and legislative framework exists, which guides policy formulation at the basin and national levels (Jacobs, 2009, pp. 100–101).

The most significant regional and basin-wide legal instruments for the OSRB include the 2000 SADC Revised Protocol on Shared Watercourses, bilateral agreements that were established between the four OSRB riparian states and one multilateral basin-wide treaty (Jacobs, 2009, p. 114; Kistin, 2012, p. 42). Raadgever *et al.* (2008, p.10) argued that international law in the basin consists primarily of the SADC revised protocol and the legal framework at the basin and national levels consist primarily of ORASECOM and bilateral agreements.

According to Kistin and Ashton (2008, p. 391) and Kistin (2012, p. 42), bilateral agreements relevant to the current management of the basin include:

- The 1986 bilateral treaty between South Africa and Lesotho, providing a framework for the LHWP and the establishment of the Joint Permanent Technical Commission (JPTC), referred to today as the Lesotho Highlands Water Commission (LHWC).
- The 1992 bilateral agreement to establish the Vioolsdrift and Noordoewer Joint Irrigation Scheme (VNJIS) and the Joint Irrigation Authority (JIA) between South Africa and Namibia.
- The 1992 bilateral agreement between South Africa and Namibia that resulted in the establishment of the Permanent Water Commission (PWC).
- The 2000 multilateral agreement establishing ORASECOM between all four riparian states.

The 1986 LHWP Treaty, a key bilateral agreement between South Africa and Lesotho, is a project-based treaty that established provisions to construct and manage the LHWP. Similarly, the 1992 VNJIS agreement addressed the planning, operation, and maintenance of joint projects in the basin specific to the VNJIS. Additionally, it dedicates 20 million m^3 annually to the scheme, with 11 million m^3 going to farmers in South Africa and 9 million m^3 designated for those in Namibia. The agreements establishing the PWC and ORASECOM create joint institutions to advise parties on the development and use of shared waters (Kistin and Ashton, 2008, pp. 391–393; Kistin, 2012, p. 42).

In terms of institutional responsibility, the project-related institutions, the LHWC and the JIA are granted substantial powers to design and carry out policies and procedures relating to the investigation, negotiation, and recommendation to parties regarding water allocation. The two commissions, for example, the PWC and ORASECOM, serve as advisory bodies whose mandates are wider in scope than the project-based institutions and were specifically designed with an advisory function to parties on *such matters as may be determined* by the parties (Kistin and Ashton, 2008, p. 396).

Regional context

Water in the region is a scarce resource. It is foreseen that in the next 20 to 30 years, three or four SADC states will be facing serious water shortages. As a result of this estimation, and in recognition of the importance of a coordinated approach to the use and preservation of water, SADC member states initiated and signed the SADC *Protocol on Shared Watercourse Systems* (Muller *et al.*, 2015, p. 51). The protocol was signed at the 1995 summit in South Africa and came into force on 29 September 1998, when two-thirds of the original SADC member states had ratified it (SADC, 2000: Article 16.1; Mbaiwa, 2004, p. 1323).

However, in Windhoek on 7 August 2000, the protocol was revised and came into force in October 2003, after Tanzania became the ninth member state to ratify it. Angola, Botswana, Lesotho, Malawi, Mauritius, Mozambique, Namibia, Seychelles, South Africa, Swaziland, Tanzania, Zambia and Zimbabwe are all signatories to the protocol (SADC, 2000, p. 17; Solomon, 2014, p. 89). The revised SADC protocol is aligned with the 1997 UN Convention on Law of the Non-Navigational Uses of International Watercourses. This is reflected in the preamble of the protocol, which states that the protocol bears,

> in mind the progress with the development and codification of international water law initiated by the Helsinki Rules and that the United Nations subsequently adopted the United Nations Convention on Law of the Non-Navigational Uses of International Watercourses.
>
> (SADC, 2000, p. 1)

This serves to align the regional protocol with internationally accepted norms on shared watercourses (Furlong, 2006, p. 448). Some of the key objectives of the protocol are the harmonisation of transboundary water policies between riparian states within transboundary river basins and seeking basin management of resources (Sebastian, 2008, p. 155).

The main thrust of the protocol is to ensure equitable sharing of water and efficient conservation of scarce resources. According to Article 2(b) of the revised protocol, the protocol seeks to:

> advance the sustainable, equitable and reasonable utilisation of the shared watercourses
>
> (SADC, 2000, Article 2b)

The revised protocol also provides guidelines to manage such shared watercourses in Article 4.3. In this regard, and upon the request of a watercourse state, states who share a watercourse should,

> enter into consultations concerning the management of a shared watercourse, which may include the establishment of a joint management mechanism
>
> (SADC, 2000, Article 4.3a)

Article 5.3(a–c) of the protocol then called for the establishment of river basin organisations or shared watercourse institutions (SWIs) (SADC, 2000: Article 5.3(a–c); Jacobs, 2009, p. 113).

In addition to calling for the establishment of river basin organisations or SWIs, Article 5.4 also calls for an *institutional framework for implementation* on the SADC level. SADC now has a fully functional water sector, composed of a number of organs, such as the Committee of Water Ministers, and the Committee of Water Senior Officials. In addition to the concept of harmonisation explicitly mentioned in Article 2(d) of the revised protocol, is the promotion of regional integration. Regional integration is one of SADC's overall objectives and is a process that requires focused cooperation, joint decision-making, and suitable institutional arrangements between states. When sharing the use of a single watercourse, such as the Orange Senqu River, integration, at least at a basin-wide level, seems unavoidable. The protocol provides the necessary framework, which should be fleshed out in a specific arrangement between the states involved (SADC, 2000, Article 2; Jacobs, 2009, p. 113).

River basin organisations or SWIs that support joint infrastructure, planning and cooperation have been established across the region in line with the protocol. The water division at the SADC Secretariat provides a platform for member states to address water-related issues and challenges (*Southern Times*, 2015).

Basin-wide context

In a basin-wide context, it is important to note that national water laws are explicitly linked to international agreements. They have undergone several adjustments and updates recently to be compatible with international agreements (Raadgever *et al.*, 2008, p. 10).

Bilateral regimes

THE LWHP

The LHWP, which consists of one phase divided into two categories, Phases 1A and B, is one of the world's largest transboundary water projects (Wirkus and Böge, 2006, pp. 26–27). Phase 1A, constructed at a cost of approximately ZAR 20 billion, included the construction of a large dam at Katse on the Malibamatso river (the highest dam in Africa at 180 m), a 45 km transfer tunnel to 'Muela hydropower station and 'Muela tail-pond, and a further 37 km delivery tunnel to the Ash River in South Africa (Hitchcock, 2015, p. 527).

Phase 1B included the Mohale Dam (the highest rock-filled dam in Africa at 145 m), and the Mohale Reservoir, located on the Senqunyane River. There is also a 32 km tunnel connecting the Mohale Reservoir to the Katse Reservoir, and a weir at Matsoku near 'Muela on the Matsoku River, from which there is a 5.6 km transfer tunnel to the Katse Reservoir. The 72 MW hydropower station at 'Muela connects with the southern African power pool and provides Lesotho with its electricity needs (Hitchcock, 2015, p. 527).

The primary funder of the LHWP, as perceived by the governments of Lesotho and South Africa, respectively, was the World Bank's International Bank for Reconstruction and Development (IBRD), which provided a loan of USD 45,000,000 to Lesotho (World Bank, 2007). According to Hitchcock (2015, p. 526), this amount represents approximately 3% of the total project cost. In addition to funding from the IBRD, the project received funding from the Government of Lesotho, the Development Bank of Southern Africa, the European Investment Bank, the African Development Bank, and various commercial banks and institutions.

There are three major reasons for both governments perceiving the IBRD to be a primary funder despite funding from other institutions. These reasons include: (1) having the World Bank on board would encourage other potential funding agencies to get involved; (2) Lesotho was on the list of the World Bank's lowest-income countries, and therefore, would qualify for a loan; and (3) both governments felt that having the World Bank support, monitor, and evaluate the project would ensure that the project followed World Bank guidelines; therefore, making it more likely to be acceptable to stockholders of companies, nongovernment organisations (NGOs), and communities concerned about social, economic, and environmental issues (Hitchcock, 2015, p. 526).

The LHWP was conceived as a means of resolving the problem that was posed by South Africa's need for a secure water supply by tapping the water-rich Senqu in Lesotho to meet the Gauteng region's water needs (Klaphake, 2005). The LHWP and its institutional framework have been amalgamated with ORASECOM.

The project is designed to divert the waters of the Senqu in Lesotho to Gauteng via a complex system of dams and tunnels and, at the same time, to generate hydroelectricity. The idea is that water transfer and power generation will benefit the development of both the countries involved. Lesotho, a country that uses only a small part of its abundant water resources, provides low-cost water to its water-poor neighbour South Africa, for power generation. Lesotho is paid royalties in return (Klaphake, 2005).

Engaging third parties or international donors (i.e., the World Bank) contributed significantly to the LHWP's success. While the role of the World Bank as a provider of credit is quite limited, the bank has financed ≤3% of the overall project costs, its engagement did encourage other donors, such as the European Investment Bank to become involved in project financing. The first phase was concluded in 2004 and, since then, water has been transferred at a rate of 29 m^3/s, which is the equivalent of roughly one-quarter of the Senqu's total flow (Wirkus and Böge, 2006, pp. 26–28).

The legal basis for the project is a bilateral government treaty that was signed in October 1986. This is one of the most comprehensive and detailed water-related agreements concluded in sub-Saharan Africa (Turton 2004, p. 274). It lays down clear-cut and binding rules that govern the behaviour and the duties of the parties (LHWP Treaty, 1986). Since then, six protocols have been added to the treaty. They may be seen as reflections of the experiences in implementation as well as of newly emerging problems, several concerning the project's institutional underpinning (Wirkus and Böge, 2006, p. 23).

Originally, each of the parties established an independent implementing organisation. In Lesotho, the LHDA is responsible for dam construction and operation (LHWP Treaty 1986, Articles 7 and 8). Although viewed as essential to the project's poverty reduction goal, the LHDA remains a weak and unsatisfactory institution (World Bank, 2007, p. 9). For instance, the World Bank (2007) report criticised the LHDA for its poorly edited reports. In South Africa, the Trans-Caledon Tunnel Authority (TCTA) is responsible for the construction and operation of the tunnel system as well as for financial and credit management (LHWP Treaty, 1986, Articles 7 and 8). The JPTC was created as a coordinating and oversight body, with three representatives from each party forming part of the Commission (LHWP Treaty, 1986, Article 9).

Since this structure proved unwieldy, a comprehensive reform was undertaken in 1999. The JPTC was reorganised to form the LHWC, a body with markedly strengthened competences and capacities (e.g., permanent secretariat and biweekly LHWC meetings). The LHDA and TCTA were restricted to a number of highly specific powers and made more subordinate to the LHWC, a move that served to streamline the organisational structure (Protocol VI, 1999).

> The LHWC has monitoring and advisory powers over the administrative, technical, and financial activities of the project
>
> (Turton et al., 2004, p. 241)

One aspect of the LHWP's institutional structure, viewed to be a special feature, is its dispute settlement mechanism. There is the Arbitral Tribunal, a dispute settlement body that is key to settling disputes in cases where a dispute emerges between the parties that cannot be settled through negotiations [LHWP Treaty, 1986, Articles 16, (15) (b)]. The tribunal is made up of three members, none of whom may be from a party to the treaty. Each side appoints one member. The members, in turn, appoint a third person as their chairman. The members of the tribunal are appointed on an ad hoc basis. If they are unable to reach an agreement, the president of the International Commission on Large Dams appoints a tribunal chairman, who may not be a citizen of either Lesotho or South Africa. The body issues binding decisions on a majority basis (Wirkus and Böge, 2006, pp. 27–28).

Furthermore, the World Bank has also influenced the project's political context. It worked to incorporate into the project minimum social and environmental standards, obtained permission for the project from SWAPO (at that time the internationally recognised representative of Namibia, then still under South African control), and made its engagement conditional on a Namibian no-objection statement (Conley and van Niekerk, 2000, p. 144 f.).

Finally, the World Bank's engagement also served to strengthen the hand of Lesotho, being the economically far weaker partner, in its dealings with its powerful neighbour (Turton, 2003, p. 147). Originally, no institutionalised stakeholder participation was provided for in the LHWP framework. Nonetheless, the World Bank's contribution to the LHWP also faced criticism from various national and international environmental and human rights NGOs (the Highlands Church and Solidarity Action Group from Lesotho and the International Rivers Network among others) (Wirkus and Böge, 2006, p. 28).

These criticisms included the environmental, social, and political impacts of the project and its lack of social and environmental compatibility. This was specifically with regard to the affected land and population (Wirkus and Böge, 2006, p. 28). These NGOs were influential in pressuring the World Bank and representatives of the governments involved, through an agreement reached in 1999 via a memorandum of understanding between the LHDA and interest groups in Lesotho, to enhance cooperation in the following areas:

> monitoring and evaluation activities; advocacy activities; community empowerment activities; service provision; and delivery activities
>
> (Meissner 2000, p. 26)

THE PWC

In September 1992, the PWC, a bilateral institution between South Africa and Namibia was created. The commission's task is to provide the governments of both parties with advice on developing the Lower Orange River (the section along which the Orange River forms the border between the two countries). The PWC replaced a Joint Technical Committee set up in 1987 as a bilateral agreement between Namibia and South Africa while Namibia was still an autonomous region of South Africa. In 1992, following Namibia's independence in 1990, a bilateral agreement between Namibia and South Africa was established which led to today's PWC (Tompkins, 2007, p. 8).

The PWC acts in the capacity of a,

> technical adviser to the Parties on matters relating to the development and utilisation of water resources of common interest to the Parties
>
> (PWC Agreement, 1992, Articles 1 and 3)

This advisory function consists mainly of the preparation of reports. The commission is made up of three delegates from each party but can appoint additional advisers. Decisions are taken based on consensus. Each of the parties is responsible for bearing the costs of its delegation, with each host country assuming the costs for the meetings held there. Nonetheless, the parties share all other costs that may accrue. A special feature of the PWC's operation is the inclusion of ORASECOM. For instance, ORASECOM is regularly informed of all PWC activities (Conley and van Niekerk, 2000, p. 141).

Disputes and differences are settled in consultations between the parties. One other aspect associated with the 1992 PWC agreement is the establishment of a JIA for a large-scale irrigation system on both sides of the Lower Orange River (e.g., Noordoewer in Namibia; Vioolsdrift in South Africa: the 800 ha VNJIS) (Heyns, 2004, p. 8). According to Van Rensburg (2014, p. 27), in 1933 when the then South West Africa was under South African control, the government constructed an irrigation scheme on both sides of the river to stimulate agricultural development and job creation. Namibian independence in 1990 meant that the scheme could no longer remain under South African jurisdiction. In 1993, the JIA was established, with a board made up of representatives of both farming communities, the South African Department of Water Affairs (now the DWS) and the Namibian Department of Agriculture.

The PWC also has an Arbitral Tribunal, which serves as a dispute settlement mechanism. Moreover, there is a separate institution that focuses on matters relating to the Orange River mouth, because the Orange River mouth is a wetland under a sustainability threat and is listed in the Montreux Record of Wetlands Under Threat. As numerous industries are central to this threat, government representatives of both parties as well as representatives of the private sector (mining) have seats on this Orange River Mouth Interim Management Committee (Heyns, 2004, p. 8).

However, the PWC has a weak organisational structure. For instance, meetings are convened from case to case, as well as *sub-committees dealing with joint irrigation and planning tasks* (Conley and van Niekerk, 2000, p. 141). Although ORASECOM is regularly informed of all PWC activities, making it obvious that it is aware of this weak organisational structure, it has failed to address the issue. It is evident that this is one of many crucial reasons for the border dispute between both parties not being resolved to date.

Multilateral regimes

ORANGE SENQU RIVER COMMISSION

An agreement was signed in 2000 by the OSRB states Botswana, Lesotho, Namibia, and South Africa to establish ORASECOM, which serves to enable the development of the Orange River for the benefit of all in the respective basin states. The organisation's founding document refers to and explicitly recognises, the Helsinki Rules, the 1997 UN Convention, and the SADC revised water protocol. ORASECOM is the first formal body established to manage shared water resources since the Protocol on Shared Watercourse Systems became an instrument of international water law in SADC (*Southern Times*, 2015). ORASECOM was established under the *2000 SADC Revised Protocol for Shared Watercourses* (Sebastian, 2008, p. 147).

ORASECOM was to create a framework for multilateral riparian cooperation to investigate and make decisions about future actions dealing with socio-economic and political issues concerning water management, sustainability, and use of the OSRB (Sebastian, 2008, p. 147). Despite this, the central problem of unequal access to water remains essentially unsolved.

ORASECOM serves as a cooperation platform for member states to address water-related issues and challenges. According to ORASECOM executive secretary, Lenka Thamae,

> One of the key values of the four ORASECOM countries – Botswana, Lesotho, Namibia and South Africa – working collectively on a river basin is that it aligns planning and forecasting/projection on future demands and resource availability. In this way, the Commission promotes co-operation beyond theory
>
> (*Southern Times*, 2015)

ORASECOM is in line with the principle that the integrated and sustainable management of international river basins should involve all riparian countries. The commission aims to develop a comprehensive perspective of the OSRB by studying the present and planned future uses of the river system and by determining the requirements for flow monitoring and flood management. It is expected to strengthen regional solidarity, contribute to peace and harmony, and enhance socioeconomic cooperation.

It is important to note that the existence of a multilateral regime does not replace existing bilateral regimes, nor does it preclude additional future bilateral projects between any of the watercourse states. Rather, it provides a broader forum for overall consultation and coordination between the watercourse states for sound IWRM and development in the basin. However, future bilateral agreements must be in line with ORASECOM.

ORASECOM is authorised to make recommendations based on joint water yield assessments, conduct appropriate feasibility studies, and develop joint assessments of water allocation and appropriate infrastructure development. Despite this, many hydro infrastructures (making specific reference to the LHWP), were criticised for a lack of proper feasibility studies.

In cases of dispute, the commission's founding document refers to the SADC Tribunal as a dispute settlement mechanism. This is according to Article 8 of the commission, which states that disputes that cannot be resolved through commission negotiations are to be brought before the SADC Tribunal, whose decisions are to be recognised as *final and binding*. The political level of the commission is authorised to make decisions in instances where cases involve differences in opinion between member states.

A major aspect central to ORASECOM's operational difficulty includes dependence on foreign aid. Finance from donors almost always comes with strings attached. Regimes are sometimes unable to align their priorities to those of their funding partners. Partners, in most instances, prefer certain areas of cooperation that differ from those of the parties directly involved. According to Raadgever *et al.* (2008, p. 10), the development of transboundary institutions has been driven by donors who have been involved in financing the establishment of the ORASECOM, financing participatory processes, and financing concrete research projects in the basin. Donor funding may not be the ideal financial source for adaptive management, despite its contributions to the development of cooperation and more complete laws (Raadgever *et al.*, 2008, p. 10). Although ORASECOM is funded by the water ministries of the countries concerned, it also receives financial support from Germany (through the German Technical Corporation), France, and the European Union (Heyns, 2004, p. 9).

However, on a more positive note, ORASECOM is gradually receiving funding from continental organisations that are starting to become financially accountable and responsive to their issues. The New Partnership for Africa's Development (NEPAD), an economic development programme of the African Union, has provided ORSECOM with ZAR 48.6 million in funding towards water security initiatives. In January 2017, the African Water Facility, and the NEPAD Infrastructure Project Preparation Facility (NEPAD–IPPF) signed an agreement with ORASECOM to launch the Climate Resilient Water Resources Investment Strategy and Multipurpose Project Preparation. The project received EUR 3.5 million (approximately ZAR 48.6 million) of which the African Water Facility and NEPAD-IPPF contributed EUR 2 million and USD 1.2 million, respectively (Naidoo, 2017).

This cofinanced project is intended to promote sustainable socioeconomic growth in the basin's riparian countries through climate-resilient water resources development. ORASECOM similarly believes that this project will foster enhanced sustainable water resources management of the basin and increased investments based on better planning, which will in turn allow for more multipurpose projects. Climate resilience must also be integrated into the planning and development of water infrastructure in the basin. This will ultimately allow ORASECOM to effectively and efficiently address the livelihood needs of the communities living in rural and urban areas (Naidoo, 2017). This, therefore, symbolises a shift in continental organisations from a dependence on foreign aid approach to a more accountable approach, meaning that enhanced water resource sustainability, management, and use is slowly becoming a reality on the African continent, particularly in the OSRB.

Given the scale of institutional development in the region; however, Lesotho faces difficulties in proper policy planning that will ultimately affect basin-wide institutional planning as far as policy harmonisation and basin management is concerned. In Lesotho, there is no coordinated approach to dealing with land and water resources. For instance, Minister Mathaba reiterated,

> Government ministries were dealing with water, land and other related natural resources issues in their individual capacities. As a result, such independent campaigns were not producing the desired results
>
> (*Lesotho Times*, 2017)

It is essential to note that land and water issues cannot be individually addressed. According to Baecher *et al.* (2000, p. 73), land issues and land degradation are inextricably linked to the quality, volumes, and timing of water flow in the basin.

Mathaba viewed sectors, such as finance, planning, agriculture, energy, tourism, industry, education, and health as crucial for integrated land and water resource management. However, he noted that such integration remains a challenge since,

> various institutions responsible for management of water, land and related resources are applying their own individualised evaluation strategies with their unique objectives and methods. Therefore the impact of these measures is equally limited.

This will further exacerbate the complexity of implementing integrated catchment management (*Lesotho Times*, 2017).

Moreover, these sentiments are shared by NEPAD–IPPF manager Shem Simuyemba, who asserts that the lack of inclusive water resource sustainability, management and use will further exacerbate the already existing multiple problems related to deteriorating environmental conditions. This will ultimately make it difficult, if not impossible, to support economic growth and alleviate poverty (Naidoo, 2017). It is therefore evident that Lesotho would need to ensure that different institutions combine their efforts in a coordinated manner to improve its land and water resources management and achieve sustainable development goals.

On a positive note, Lesotho's water affairs ministry has adopted various legal tools, such as the Water and Sanitation Policy of 2007, the Water Act of 2008, and the Long-Term Water and Sanitation Strategy in 2016 as part of its efforts to address the situation (*Lesotho Times*, 2017).

Conclusion

Although there is progress as far as policy developments in the OSRB are concerned, poor policy implementation remains. For instance, water management in the basin has traditionally concentrated on large-scale infrastructures, such as dams and water transfer pipelines, tailored towards meeting the short-term water demands of individual countries. However, there are long-term adverse effects of large-scale infrastructure (Raadgever *et al.*, 2008, p. 11).

One of the problems central to policy formulation, implementation, or both is a dependence on foreign aid and the failure of member states to take financial accountability for their institutions. Although it can be argued that the financial contributions of international donors have been quite instrumental in developing large infrastructural works, which increased the availability of resources, this also increases dependence on third parties. The OSRB financing system is still dependent on foreign donors; therefore, compromising institutional capacity and effectiveness (Raadgever *et al.*, 2008, p. 11).

For instance, from a regional and basin-wide perspective, SADC and the PWC have failed to mediate the border dispute between South Africa and Namibia, which has been going on for years and remains a problematic issue. The problematic aspect of this border dispute is that it is one of the oldest boundary disputes in the world; therefore, further questioning the effectiveness of relevant institutions that govern the basin (Oduntan, 2015). Oduntan (2015) further noted that border disputes have been a reality on the continent throughout the millennia. Today, close to 100 active border disputes exist across the continent. Rising nationalism, population, and environmental pressures mean that the situation is likely to get worse.

In general, regarding a link between a nascent environmental conflict and regime capacity in the OSRB, this chapter can draw two important findings. First, the increased environmental scarcity evident in the OSRB will lead to further resource capture and extensive ecological marginalisation. This will subsequently challenge the authority and effectiveness of governing water institutions and possibly weaken them, collapse them, or both. As a result, the OSRB riparian states and societies will be exposed to the effects of environmental scarcity and, due to this vulnerability, will fail to adapt. In the same process of vulnerability is the possible emergence of localised and regional environmental conflicts where stronger groups, states, or both (South Africa in this case) will attempt to confiscate large proportions of the shared resource and further restrict access to the basin by other parties (Namibia and Lesotho in this case) through treaties.

As witnessed in the case of Lesotho and South Africa and South Africa and Namibia, their behaviour isn't converging because of agreements and property rights that perpetuate unequal access to, and distribution of, the OSRB. Despite the existence of levels of sharing information, bilateral and multilateral agreements, and harmonised national laws to pursue common standards in the basin, such structurally induced scarcity will further exacerbate the already existing mistrust and fear of exploitation among the OSRB riparian countries.

However, the OSRB riparian countries can use the already functional and existing regimes to their advantage and adapt to the effects of environmental scarcity; therefore, avoiding a nascent environmental conflict. This is only possible under certain circumstances. First, a sense of strong political will is required of the OSRB water regimes, which will subsequently ensure institutional capacity. In light of these regimes having failed to address issues, such as land degradation caused by industries, mining and other human-induced factors, the implications of the construction of large dams and IBTs on people and the environment, and the unequal distribution of the OSRB, this chapter, therefore, deduces that a much-needed political will in the basin is lacking.

Given that political will is entrenched in governing water regimes, this will enhance institutional capacity to prevent the possible weakening and collapse of institutions. Furthermore, the OSRB riparian countries need to commit to sharing information, establishing agreements that provide for equitable access and distribution, and harmonising and updating national laws to combat the effects of land degradation (all of which are essential ingredients for establishing common standards). The results of these commitments will lay the foundation for regimes to address environmental scarcity and therefore develop effective dispute resolution processes meant to address potential environmental conflict, either local or regional.

The task of securing a sufficient supply of water for the economic development of the OSRB will remain a challenge in the future. Despite the LHWP solving Gauteng's water scarcity problems while, at the same time, benefiting Lesotho in economic terms, it is important not to ignore the ecological, social, and political costs involved (Wirkus and Böge, 2006, pp. 26–27). Furthermore, attempts to push through further phases of the LHWP could well entail further social and environmental risks. The problem of water quantity will soon be exacerbated by an increasingly pressing quality problem due to water pollution. Conflicts may develop between South Africa and Namibia over the development of the Lower Orange River.

To address the tensions and disputes between Lesotho and South Africa over the LHWP Treaty, a rectification of the treaty is very necessary. In addition, this implies adjustments to the dates of review of the treaty to kerb the current climatic pressures. After all, Paragraph 1, Article 16 of the treaty titled *Procedure for Review and Revision* states,

> The provisions of this Treaty shall be reviewed at intervals of twelve years, calculated from the date of signature hereof or at such other intervals as the Parties may agree upon.
>
> The third of such intervals would be the year 2022
>
> (News24, 2016)

It is evident that bilateral agreements and multilateral treaties have dominated co-riparian relations in the OSRB (Kistin and Ashton, 2008, p. 396). According to Jacobs (2009, pp. 113–114), institutional development on the OSRB has been fragmented but successful where it has occurred, reaching a level of sophistication and success not found in other river basins in southern Africa. Moreover, institutional arrangements have evolved and reflected the changing political, social, and economic transformations that have occurred in the region (Kistin and Ashton, 2008, p. 391). However, the evolution of these institutions, as witnessed earlier, is still confronted by the challenge of foreign aid dependence. Furthermore, these institutions have failed to resolve critical issues in the basin, specifically, disputes over borders transcending through transboundary basins.

Despite the South Africa–Namibia border dispute not yet culminating in an outright transboundary conflict, it poses a serious threat to the stability of the region. It also questions the effectiveness of the SADC to take accountability and deal with nascent transboundary disputes in the region, particularly in the OSRB. This is also reflective of the dispute between Namibia and Botswana over the Okavango River Basin, which was resolved by the International Court of Justice (ICJ) instead of the regional body. This book also found that Namibian exploitation of the Okavango River was a clear source of disagreement with Botswana.

With about 13 contentious cases concerning territorial or boundary disputes between states being submitted to the ICJ and the Permanent Court of Arbitration, it is evident that African states have been making use of, and relying heavily on, these international judicial institutions (Oduntan, 2015).

The inability of SADC to take responsibility for territorial disputes emerging within its jurisdictions means that cases will be confronted by Eurocentric international law (seeing that the compositions of both judicial institutions are unrepresentative of Africa) rather than regional or continental-based law; therefore, compromising the interests of African countries. The only instance when African states initiate resolving border disputes is using indigenous mechanisms; therefore, questioning the legitimacy of those mechanisms. For instance, to resolve the Namibia–South Africa border dispute over the OSRB, the then presidents of Namibia and South Africa, Sam Nujoma and Nelson Mandela, respectively, only managed to reach *a gentlemen's agreement*. Maybe this is a sufficient reason for the bone of contention over the border persisting even today.

ORASECOM's weak capacity led to the agreement that support from foreign donors is still crucial at this point to stabilise and monitor the progress of the commission. Sources suggest that the German DC should continue to support ORASECOM (assistance in setting up a secretariat and in working out an IWRM plan) and undertake to ensure that the positive experiences with ORASECOM are disseminated and used in other river basins (ORASECOM as the core of a centre of competence).

In conclusion, it is important to note that environmental stresses (e.g., rapid population growth, resource degradation and depletion, resource distribution, and climatic variations because of climate change) and regimes (e.g., institutions, treaties, and agreements) influence each other. Both these aspects determine the emergence and extent of environmental conflicts. As Earle (2005, p. 57) stated earlier, the institutional capacity to adapt, address, or both environmental stress is inherent to the emergence of environmental conflicts (Earle, 2005, p. 57). Similarly, Hendrix and Salehyan (2010) noted that the interaction of environmental stresses with ideational factors, such as institutions, regimes, or both is key to whether water-related conflicts develop. Because capacity building supports the ability to manage water resources effectively and to maintain local water infrastructure, it ultimately contributes to riparian countries' and people's resilience (Meyer, 2013, p. 57). Despite the conflict potential in the OSRB as a result of existing disagreements, the basin's high level of institutionalised cooperation and the possibilities for IBTs could help to alleviate water stress and resolve bilateral disagreements over shared water resources.

Notes

1 Area of land formed by Vereeniging, Vanderbijlpark and Sasolburg – together they comprise a substantial urban complex in South Africa about 60 km south of Johannesburg.

Meyerton, just north of Vereeniging, is also generally included in the complex, and residents of Sharpeville, Boipatong, Bophelong, the greater Sebokeng area (e.g., Evaton and Orange Farm), Heidelberg, Zamdela, and Potchefstroom
2 Availability of ≤1,000 m^3 per person/year of freshwater.
3 Availability of ≤1,000–1,700 m^3 per person/year of freshwater.
4 A form of irrigation that saves water and fertiliser by allowing water to drip slowly to the roots of plants, either onto the soil surface or directly onto the root zone, through a network of valves, pipes, tubing and emitters (Brouwer *et al.*, 2001, p. 52).

Bibliography

Baecher, G. *et al.* (2000) 'The Nile basin: Environmental transboundary opportunities and constraints analysis', United States Agency for International Development Draft Paper, October.Washington, D.C: United States Agency for International Development.

Bennett, O. (2012) 'The human cost of displacement', *Panos London*, 14 May. Available at: http://panoslondon.panosnetwork.org/countries/lesotho/. (Accessed: 12 October 2016).

Brouwer, C., Prins, K., Kay, M. and Heibloem, M. (2001) *Irrigation water management: Irrigation methods*. FAO Land and Water Development Division. The International Institute for Land Reclamation and Improvement, Wageningen, The Netherlands.

Burnett, P. (2009) 'Southern Africa: Neglected land washing away', *Inter Press Service*, 30 September, Available at: www.ipsnews.net/2009/09/southern-africa-neglected-land-washing-away/. (Accessed: 11 August 2016).

CIA World Factbook. (n.d) 'Disputes – International',Available at: https://www.cia.gov/the-world-factbook/field/disputes-international/. (Accessed: 24 April 2016).

CIA World Factbook. (n.d.) 'Lesotho', Available at: https://www.cia.gov/the-world-factbook/countries/lesotho/. (Accessed: 12 October 2016).

Conley, A.H and Van Niekerk, P.H. (2000) *Sustainable management of international waters: The Orange River case*, Water Policy Vol.2, Issues 1-2. Amsterdam: Elsevier.

Diederichs, N. *et al.* (2005) 'Orange River Basin baseline assessment report', *NeWater - New approaches to adaptive water management under uncertainty*. Stockholm: Stockholm Environment Institute.

Earle, A. (2005) 'Hydropolitics in Southern Africa: What is the prognosis for peaceful development of shared watercourses?', Forschungsinstitut Wasser und Umwelt (FWU)/ University of Siegen Research Institute for Water and Environment FWU, Vol.3, Topics of integrated Watershed Management – Proceedings, Summer School. University of Siegen Research Institute for Water and Environment. Siegen.

Earle, A., Malzbender, D., Turton, A., and Manzungu, E. (2005) *A preliminary basin profile of the Orange Senqu Rive'*. Pretoria: Centre for International Political Studies (CIPS), University of Pretoria.

Furlong, K. (2006) 'Hidden theories, troubled waters: International relations, the 'territorial trap', and the Southern African development community's transboundary waters', *Political Geography*, 2(4).

Hasenclever, A., Mayer, P., and Rittberger, V. (2000) 'Integrating theories of international regimes', *Review of International Studies*, 26.

Henwood, R. and Funke, N. (2002) 'Managing shared watercourse systems in Southern Africa: International relations or foreign policy?' in Turton, A. and Henwwood, R. (eds.)

Hydro politics in the developing World: A Southern African perspective. Pretoria: African Water Issue Research Unit, pp. 177-186.

Hepworth, N.D. (2009). *A progressive critique of IWRM in sub-Saharan Africa: Beyond capacity towards self-determined regulatory personality*. Norwich University of East Anglia.

Heyns, P, Patrick, M. and Turton, A. (2008) 'Transboundary water resource management in Southern Africa: Meeting the challenge of joint planning and management in the Orange River Basin', *International Journal of Water Resources Development*, 24(3).

Heyns, P. (2004) 'Achievements of Orange–Senqu River Commission in integrated transboundary water resources management', *General assembly of the international network of basin organisations*, 24–28 January. Martinique. Ministry of Agriculture, Water and Rural Development.

Heyns, P. (2003). 'Water-resources management in Southern Africa', in Nakayama, M. (ed.) *International waters in southern Africa*. Tokyo: United Nations University Press, pp. 5-37.

Hiddema, U. and G. Erasmus. (2007) *Orange River Integrated Water Resources Management Plan. Legislation and legal issues surrounding the Orange River Catchment report*. ORASECOM. Gauteng: Centurion.

Hitchcock, R.K. (2015) 'The Lesotho Highlands Water Project: Dams, development, and the World Bank', *Sociology and Anthropology*, 3(10).

Homer-Dixon, T.F (1991) 'On the threshold: Environmental changes as causes of acute conflict', *International Security*, 16(2).

Homer-Dixon, T.F. (1994) 'Environmental scarcities and violent conflict: Evidence from cases', *International Security*, 19(1).

Homer-Dixon, T.F. (1999a). *Environment, scarcity, and violence*. New Jersey: Princeton University Press.

Homer-Dixon, T.F (1999b). Threshold of turmoil: environmental scarcities and violent conflict', in Deudney, D.H. and Matthew, R.A. (eds.) *Contested grounds: Security and conflict in the new environmental politics*. Albany: State University of New York Press.

Jacobs, I.M. (2009). *Norms and transboundary co-operation in Africa: the cases of the Orange-Senqu and Nile rivers*. Unpublished PhD thesis. University of St. Andrews.

Kingdom of Lesotho and Republic of South Africa. (1986) Treaty on the Lesotho Highlands Water Project between Lesotho and South Africa. Available at: www.lhda.org.ls/lhda web/uploads/documents/governance/lhwp_treaty.pdf. (Accessed: 17 October 2016).

Kistin, E. and Ashton, P. (2008) 'Adapting to change in transboundary rivers: An analysis of treaty flexibility on the Orange-Senqu River Basin', *International Journal of Water Resources Development*, 24(3).

Kistin, E. (2010) *Critiquing cooperation: The dynamic effects of transboundary water regimes*. Unpublished PhD thesis University of Oxford.

Kistin, E.J. (2012) 'Critiquing cooperation: Transboundary water governance and adaptive capacity in the Orange-Senqu Basin', *Journal of Contemporary Water Research and Education*, 149.

Klaphake, A. (2005) *Kooperation an internationalen Flüssen aus ökonomischer Perspektive: Das Konzept des Benefit Sharing, BMZ-Forschungsvorhaben, Grenzüberschreitendes Wassermanagement in Afrika*. Bonn: Deutsches Institut für Entwicklungspolitik.

Kranz, N., Interwies, E., and Vidaurre, R. (2005) Transboundary regimes in the Orange-Senqu Basin, contribution to deliverable 1.3.1 of the NeWater Project, Berlin: Ecologic – Institute for International and European Environmental Policy:

Kreamer, D.K. (2012) 'The past, present, and future of water conflict and international security', *Journal of Contemporary Water Research and Education*, 12(149).

Langea, G., Mungatanab, E., and Hassanb, R. (2007) 'Water accounting for the Orange River Basin: An economic perspective on managing a transboundary resource', *Ecological Economic*, 61.

Lee, S. (2001) 'Emerging threats to international security: Environment, refugees, and conflict', *Journal of International and Area Studies*, 8(1).

LHWP Treaty. (1986). Treaty on the Lesotho highlands water project between the government of the Kingdom of Lesotho and the government of the Republic of South Africa, 24 October. Available at: www.fao.org/docrep/w7414b/w7414b0w.htm. (Accessed: 11 November 2016).

Lesotho Times. (2017) 'Water sources under threat', 25 February. Available at: http://lestimes.com/water-sources-under-threat/. (Accessed: 13 March 2017).

Likoti, F.J. (2006) *African military intervention in African conflicts: An analysis of military intervention in Rwanda, the DRC and Lesotho*. Unpublished PhD thesis. School of Government, University of the Western Cape.

Mbaiwa, J.E. (2004) 'Causes and possible solutions to water resource conflicts in the Okavango River Basin: The case of Angola, Namibia and Botswana', *Physics and Chemistry of the Earth, Parts A/B/C*, 29(15).

Meissner, R. (2000) 'The case of the Orange River', *Conflict Trends* , 2.

Meyer, C. (2013) *Integrated water resources management – The Orange-Senqu River Basin in South Africa*. Unpublished MSc thesis. University of Hamburg.

Mokorosi, P.S. (2007) 'Can benefits be shared equitably among a variety of stakeholders in a transboundary river basin? Learning from Zambezi and Orange-Senqu River Basins', *8th WaterNet/WARFSA/GWP-SA Symposium*, 30 October to 3 November, Livingstone.

Muller, M., Chikozho, C. and Hollingworth, B. (2015) 'Water and regional integration: The role of water as a driver of regional economic integration in Southern Africa', Water Research Commission. Report No 2252/1/14, March. South Africa.

Mwangi, O.G. (2007) 'Hydro politics, ecocide and human security in Lesotho: A case study of the Lesotho Highlands Water Project', *Journal of Southern African Studies*, 33(1).

Mwangi, O.G. (2008) 'Environmental change and human security in Lesotho: The role of the Lesotho Highlands Water Project in environmental degradation', *African Security Review*. 17(3).

Mwendera, E., and Atyosi, Y. (2018) 'A review of water storage for socioeconomic development in South Africa', *Journal of Water Resource and Protection*, 10(3).

Naidoo, R. (2017) 'Orange-Senqu River Basin gets R48.6m for water security', *Infrastructurene*. 20 February. Available at: www.infrastructurene.ws/2017/02/20/orange-senqu-river-basin-gets-r48-6m-for-water-security/#. (Accessed: 14 March 2017).

Neethling, T. (1999) 'Military intervention in Lesotho: Perspectives on Operation Boleas and Beyond', *Online Journal of Peace and Conflict Resolution*, 5(2.2).

News24. (2016) 'The beauty of Lesotho's water crisis', 11 January. Available at: www.news24.com/MyNews24/the-beauty-of-lesothos-water-crisis-20160111. (Accessed: 23 August 2016.)

Oduntan, G. (2015) 'Africa's border disputes are set to rise – but there are ways to stop them', *The Conversation*, 14 July, Available at: http://theconversation.com/africas-border-disputes-are-set-to-rise-but-there-are-ways-to-stop-them-44264. (Accessed: 28 September 2016).

ORASECOM. (2008) 'Orange-Senqu River Basin: Preliminary transboundary diagnostic analysis', UNDP Main Report. Pretoria: Orange-Senqu River Commission: Centurion.

Percival, V. and Homer-Dixon, T. (1998) 'Environmental scarcity and violent conflict: the case of South Africa', *Journal of Peace Research*, 35(3).

PWC. (1992) 'Agreement between the government of the republic of Namibia and the government of the republic of South Africa on the establishment of a permanent water commission signed at Noordoewer, 14 September 1992'.

Poolman, J. (2014) 'South border dispute', Informante, 30 October. Available at: www.informante.web.na/south-border-dispute.15089. (Accessed: 28 September 2016).

Raadgever, G.T. *et al.* (2008) 'Assessing management regimes in transboundary river basins: Do they support adaptive management?' *Ecology and Science*, 13(1).

Salehyan, I., and Hendrix, C.S. (2010) 'After the rain: Rainfall variability, hydro-meteorological disasters, and social conflict in Africa', *International Studies Association Annual Conference*. Montreal. Elsevier: Amsterdam, Netherlands.

Sebastian, A.G. (2008) *Transboundary water politics: Conflict, cooperation and shadows of the past in the Okavango and Orange River Basins of Southern Africa*. Unpublished dissertation. University of Maryland.

Shilomboleni, A. (2006) 'Equitable water allocation: A case between South Africa and Namibia on the Lower Orange River', Windhoek: Department of Water Affairs and Forestry.

Solomon, H. (2014) 'Potential for cooperation rather than conflict in the face of water degradation: The cases of the Nile River and Okavango River Basins', , *Southern Journal for Contemporary History* , 39(1).

South African DWS. (2014) 'Business case for the Vaal River Catchment Management Agency (CMA)', August, Available at: www.dws.gov.za/IO/Docs/CMA/CMA/Vaal%20River%20CMA%20Business%20Case.pdf. (Accessed: 11 August 2016).

South African DWS. (2015) 'ORASECOM undertakes the second Joint Basin Survey', 23 July, Available at: www.dwa.gov.za/ Communications/PressReleases/2015/Media%20Statement-ORASECOM.pdf. (Accessed: 11 August 2016).

South African DWS. (2016) 'Water resources of the Orange River', Available at: www.dwa.gov.za/orange/waterres.aspx. (Accessed: 20 October 2016).

SADC. (2000) The revised Protocol on Shared Watercourses in the Southern African Development Community, Windhoek. Available at: https://wis.orasecom.org/content/study/ORASECOM/general/Documents/Legal/SADC_Revised_Protocol.pdf. (Accessed: 24 September 2016).

Southern Times. (2015) 'Getting water co-operation to flow', 26 October. Available at: http://southernafrican.news/2015/10/26/getting-water-co-operation-to-flow/. (Accessed: 11 August 2016).

Sullivan, C.A. (2014) 'Challenges for management of the Orange/Senqu', in Squires, V.R., Milner, H.M., and Daniell, K.A. (eds.) *River basin management in the twenty-first century: Understanding people and place*. London: CRC Press, pp. 189-215.

Swain, A. and Krampe, F. (2011) Transboundary rivers and climate change: African and Asian Rivers, in Gounden, V. (ed.) *Environment, climate change and conflict, Conflict Trends*, Issue 2. Durban: ACCORD, pp. 16-21.

Tadesse, D. (2010) 'The impact of climate change in Africa,' *Institute for Security Studies*, November. Available at: https://www.files.ethz.ch/isn/136704/PAPER220.pdf. (Accessed: 24 June 2016).

Tompkins, R. (2007) 'Institutional structures in the four Orange Basin States'. *Orange River Integrated Water Resources Management Plan*. Pretoria: WRP Consulting Engineers.

TRC. (2003) 'Resettled communities speak out', *Work for Justice*, 65.

Turton, A.R. (2003) 'An overview of the hydropolitical dynamics of the Orange River basin', in Nakayama, M. (ed.) *International Waters in Southern Africa*. Tokyo: United Nations University Press, pp. 136-163.

Turton, A.R. (2004) 'The evolution of water management institutions in selected southern African international river basins', in Biswas, K.A. Ünver, O. and Tortajada, C. (eds.) *Water as a focus for regional development*. New Delhi: Oxford University Press, pp. 251-289.

Turton, A.R., Meissner, R., Mampane, P.M., and Seremo, O. (2004) 'A hydropolitical history of South Africa's international river basins', *Water Research Commission (WRC)*, report no.1220/1/04, University of Pretoria: African Water Issues Research Unit (AWIRU).

Van Rensburg, D. (2014) 'South Africa and Namibia ensuring enough irrigation water together', *The Water Wheel*, 13(6).

Van Schaik, L. and Dinnissen, R. (2014) 'Terra Incognita: Land degradation as underestimated threat amplifier', Clingendael Report for the Netherlands Environmental Assessment Agency (PBL), Wassenaar: Netherlands Institute of International Relations:

Veilleux, J.C. (2013) 'The human security dimensions of dam development: The grand Ethiopian Renaissance Dam', *Global Dialogue*, 15(2.).

Wirkus, L. and Böge, V. (2006) 'Transboundary water management on Africa's international rivers and lakes: current state and experiences', in *Transboundary water management in Africa: Challenges for development cooperation*. Bonn: German Development Institute (DIE), pp. 11-102.

Wolf, A.T. (2005) 'Hydropolitical vulnerability and resilience: Series Introduction', *Hydropolitical vulnerability and resilience along international waters: Africa*. Nairobi: United Nations Environment Programme.

World Bank. (2007) *Supervision Mission Report*. Lesotho

Worldometers. (2016) *Southern Africa Population 2016*. Available at: www.worldometers.info/world-population/southern-africa-population/. (Accessed: 12 October 2016).

Young, O.R. (2011) 'Effectiveness of international environmental regimes: Existing knowledge, cutting-edge themes, and research strategies', *Proceedings of the National Academy of Sciences of the United States of America*, 108(50).

Zabarenko, D. (2011) 'Water use rising faster than world population', *Reuters*, 25 October. Available at: www.reuters.com/journalists/deborah-zabarenko. (Accessed: 11 August 2016).

4 The Nile River Basin

The Nile is one of Africa's most complex cross-border river basins (Ashton, 2002, p. 1), given the volatility and proximity of the basin. As a result, three distinctive but related challenges are visible. According to Kieyah (2007, p. 2),

> The Nile Basin faces considerable challenges. These challenges include rapid population growth, environmental degradation, unequal distribution and insecurity.

These challenges combined have affected the management, sustainability, and use of the basin.

First, simply because of the number of riparian states dependent on its resources, the Nile is confronted with a huge problem of rapid population growth, which ultimately places pressure on the land and water resources of the basin. Second, the Nile River's water is limited while the needs of the riparian countries continue to grow, making the potential for conflict very real. Finally, the Nile River Basin (NRB) is challenged by disputes over the unequal use of water between the upstream and downstream riparian countries (Jacobs, 2006, p. 13; Solomon, 1996, p. 3).

Shinn (2006) argued that the most important of these 11 riparian countries are Egypt and Sudan. The remaining nine riparian countries do not benefit equally from the Nile water. This is the result of two separate agreements: the 1929 Nile Agreements were concluded between Great Britain, Northern Ireland, and Egypt, and the 1959 Nile Agreement was concluded between Egypt and Sudan. In short, both agreements grant Egypt and Sudan the absolute right to use 90% of the river's water, leaving the upstream countries to share 10%, which created an unsustainable situation. Riparian countries' use, or rather the restriction of their use of the Nile, has direct implications for their national interests (Adar, 2007, p. 63). For instance, most upstream countries have a limited ability to develop economically and socially, which has resulted in political instability, famine and chronic malnutrition (El-Fadel *et al.*, 2003, p. 113). Upstream countries' developments are impaired as they are forced to either abandon certain projects due to their limited access to water or to gain approval from Egypt to use the Nile to construct water-related projects (Kameri-Mbote, 2007, p. 3).

DOI: 10.4324/9781003366379-4

A delicate conflict over the equitable allocation of the Nile's water represents one of the most crucial issues on the African continent. In parts of northeast Africa, politics have been shaped and affected by the Nile, creating a new landscape of global conflict (El-Fadel *et al.*, 2003, p. 113). Reyskens (2011) asserted that countries become susceptible to conflict if the resource they are dependent on is either threatened or removed. Such scarcities and inequalities regarding a shared resource make disputes increasingly heated (Klare, 2002, p. 139). According to Lowi (1995, p. 124), this has created a situation that could lead to an outbreak of war. It is evident that the NRB states need to establish, put into effect, or both a cooperative framework to manage, sustain, and use the basin.

Description of the NRB

The NRB, located in the region of northeast Africa, extensively plays a central role in the conflict over water. This book focuses on the NRB due to its volatility and proximity. The Nile is an international river shared by 11 riparian countries (Figure 4.1). These are Burundi, Egypt, Eritrea, Ethiopia, Kenya, Rwanda, Tanzania, Uganda, the Democratic Republic of Congo (DRC), Sudan, and South Sudan. This makes it the world's longest river at about 6,700 km or 4,100 mi (Kameri-Mbote, 2007, p. 1).

Figure 4.1 Map of the NRB.

The Nile flows across and around all 11 riparian countries, making them highly dependent on its water (Alcamo *et al.*, 1996, p. 336). These riparian countries can be divided into upstream (e.g., Burundi, Eritrea, Ethiopia, Kenya, Rwanda, Tanzania, Uganda, and the DRC) and downstream countries (e.g., Egypt, Sudan, and South Sudan) (Marterns, 2011, p. 1). Given the background of the basin, the following section unpacks the prevalent environmental scarcity challenges in the basin and their impact on basin states.

An analysis of environmental scarcity in the NRB

Population growth affecting water resources in the NRB

According to Savanije and van der Zaag (2000, p. 14), water-scarce countries that have high rates of population growth frequently find themselves involved in intra and interstate conflicts over scarce water resources. This increase in the demand for water resources generates a number of problems facing many international rivers (Marty, 2001, p. 14).

According to Baecher *et al.* (2000, p.11), the countries along the Nile are confronted with population growth. They argued that continued population increases will certainly add to the future water demands in the basin. The use of the Nile for a growing population can, therefore, become a source of conflict. The Nile Basin Initiative (NBI) (2012a) points out that population growth in the NRB is increasing at an unprecedented rate. In 2010, the total population of the basin countries was 424 million, of which 232 million lived in the basin. The initiative also estimated that by 2025, the inland parts of the basin countries' population would grow to 600 million people and more than 300 million people will live along the NRB.

The United Nations Population Division (UNPD) estimated that the total population in the NRB would reach 647 million by 2030, which represents an increase of 7.8% from the population in 2025 and 52% from the population in 2010. It is also estimated that just over half of these people will be living within the basin boundaries (NBI, 2012b, p. 228). Population growth is concentrated in the NRB and migration to the NRB has exacerbated population activity. This rapid population growth increases pressure on the natural resource base (UNEP, 2006, p. 1).

A UN report released in 2011, *Information Products for Nile Basin Water Resource Management*, also found these population projections alarming and called for new ways to use water to support better planning and prevent food insecurity. The report further argued that these population estimates are of particular concern since some of the 11 countries that share the Nile are among the poorest in the world (UN, 2011, p. 116).

As the population in the basin is expected to continue growing in the coming years, societies living close to the world's longest river are confronted by pressing social, environmental, and economic challenges (United Nations Environment Programme (UNEP), 2006, pp. 1–5). Baecher *et al.* (2000, p. 12) indicated that urbanisation will also pose special challenges as most NRB countries are expected to increase their urban populations dramatically. Challenges posed by urbanisation are related to the demand for water supply such as water delivery, drinking water, sanitation, and wastewater treatment.

Demand-induced scarcity caused by population growth is worsening. This is exacerbated by the fact that per capita water availability in the NRB countries is expected to fall dramatically (Okbazaghi, 2008, p. 5). With the predicted fall in per capita water availability, disputes over water supplies will become a serious cause of conflict among these riparian countries. For instance, it is predicted that by 2025 Egypt, Ethiopia, Kenya, Rwanda, and Burundi will experience water scarcity, while Uganda, Tanzania, and Eritrea will be under water stress (Karyabwite, 2000, pp. 6–7).

Xercavins (1999, p. 158) asserted that water scarcity is a condition in which the annual availability of internal, renewable fresh water is ≤1,000 m^3 per person. Water stress is a condition in which the annual availability of internal, renewable fresh water is <1,667 and >1,000 m^3 per person in the population.

Water scarcity is the point at which the aggregate impact of all users impinges on the supply or quality of water under prevailing institutional arrangements, to the extent that the demand by all sectors, including the environment, cannot be fully satisfied. Water scarcity is a relative concept and can occur at any level of supply or demand. Scarcity may be a social construct (a product of affluence, expectations, and customary behaviours) or the consequence of altered supply patterns stemming from, for instance, climate change (United Nations Department of Economic and Social Affairs (UNDESA), 2013).

According to the UNEP (2012, p. 18), the concept of water stress applies to situations where there is not enough water for all uses, whether agricultural, industrial, or domestic use. Based on these criteria, the UN estimates that by 2025, 1.8 billion people will be living in countries or regions with absolute water scarcity and two-thirds of the world's population could be under conditions of water stress. Due to these population projections, Stoa (2014, p. 1353) asserted that it is not surprising that the NRB is, therefore, viewed as one of the most degraded in the world. Other causes of this degradation include natural disasters, political instability, and poor watershed management, which makes the situation worse when they interact with rapid population growth.

A fall in per capita water availability coupled with rapid population growth leaves most of the riparian countries with internal and interstate conflicts, immense droughts and famine (Okbazaghi, 2008, p. 5). According to Stoa (2014, p. 1353), 70% of the NRB's population relies on rain-fed agriculture for their livelihoods. As a result, there has been a reduction in soil fertility and dry season flows, while droughts and floods put vulnerable populations at further risk of food insecurity.

Ethiopia, despite being the source of the Blue Nile, faced worsening drought conditions and famine between 1965 and 2006, making it highly dependent on international food aid (Okbazaghi, 2008, p. 5). Adding to the fall in per capita water availability are the agricultural activities that feed the growing population, which increases nonpoint pollution, siltation, and erosion (Baecher *et al.*, 2000, p. 11). The extent of water scarcity in the region has pushed the Nile's water from being a low policy priority to a top national policy priority (Okbazaghi, 2008, p. 5). The estimated growth in the population of the basin countries will place immense pressure on the Nile and its resources leading to severe degradation and negatively affecting, reducing, or both the outputs (Okascha, 2012, p. 15).

What has also worsened the issue of demand-induced scarcity in the NRB is the involvement of China in the hydropolitical landscape of the Nile, which symbolises a monumental change to NRB hydropolitics. This is witnessed through China's financial role in support of hydraulic infrastructure in upstream countries (Figures 4.2 and 4 3) (Cascão, 2009, pp. 249–251).

It is sufficient to argue that countries currently developing water resources through dam projects are doing so in response to pressures from population growth and economic development (Veilleux, 2013, p. 1). However, the social and environmental implications for the inhabitants associated with these projects are far greater than the hydropower potential expected.

These implications can be proven to be detrimental to the welfare of the Nile, with a subsequent emergence of conflict. Environmental scarcity has been exacerbated and population growth has not been addressed, resulting in thousands of displaced people in the basin who are populating small unsustainable areas.

The development of hydropower projects, given that they are developed unilaterally, does not entail equitable use; rather it follows every man for himself scenario, which will not help the peace and cooperation process in the basin. Considering that tensions have always been high between upstream and downstream countries, this further exacerbates existing delicate relations.

Furthermore, the rising demand to unilaterally develop large hydropower projects in the upstream countries has rapidly increased the consumption of the Nile and has negatively affected the supply of limited water resources that are available to each riparian country. In response to alarming power shortages, upstream countries have opted to develop large hydropower projects. However, these hydropower stations have not reversed the situation.

For instance, with a growing economy and an increase in population in Uganda, two hydropower stations (e.g., the Karuma hydropower project and the Ayago north and south hydropower dams) have not reversed the power shortage, crisis, or both and poverty in the country. These hydropower stations only meet a fraction of Uganda's electricity demand. According to the Ugandan State Minister of Energy Simon D'Ujanda, only 9% of Ugandans have electricity (Afedraru, 2014).

These projects have also been responsible for displacing communities living along the Nile. According to Pottinger (1996), these large projects have not reversed poverty and power shortages, but have made it worse through major displacements and the destruction of arable land (Pottinger, 1996). This is due to China not imposing environmental, human rights, or good governance conditions before funding (Thakur, 2009, p. 10). Pottinger (1996) further stated that the benefits of these projects were concentrated in the hands of a very small elite.

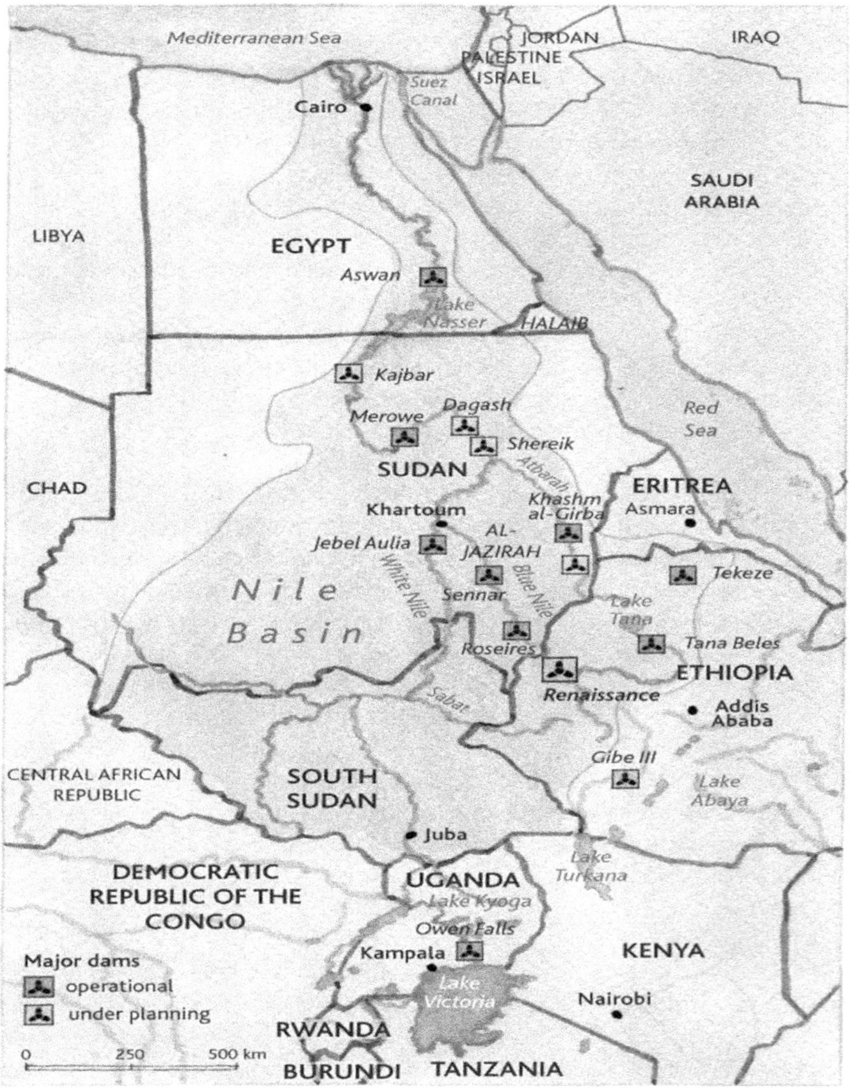

AGNÈS STIENNE

Figure 4.2 Chinese support for hydraulic projects in the Nile.

The lack of proper coordination between the Nile riparian countries has led to the construction of these hydropower projects being subject to inadequate environmental and social assessments, leaving devastating environmental and social impacts as discussed through demand-induced scarcity.

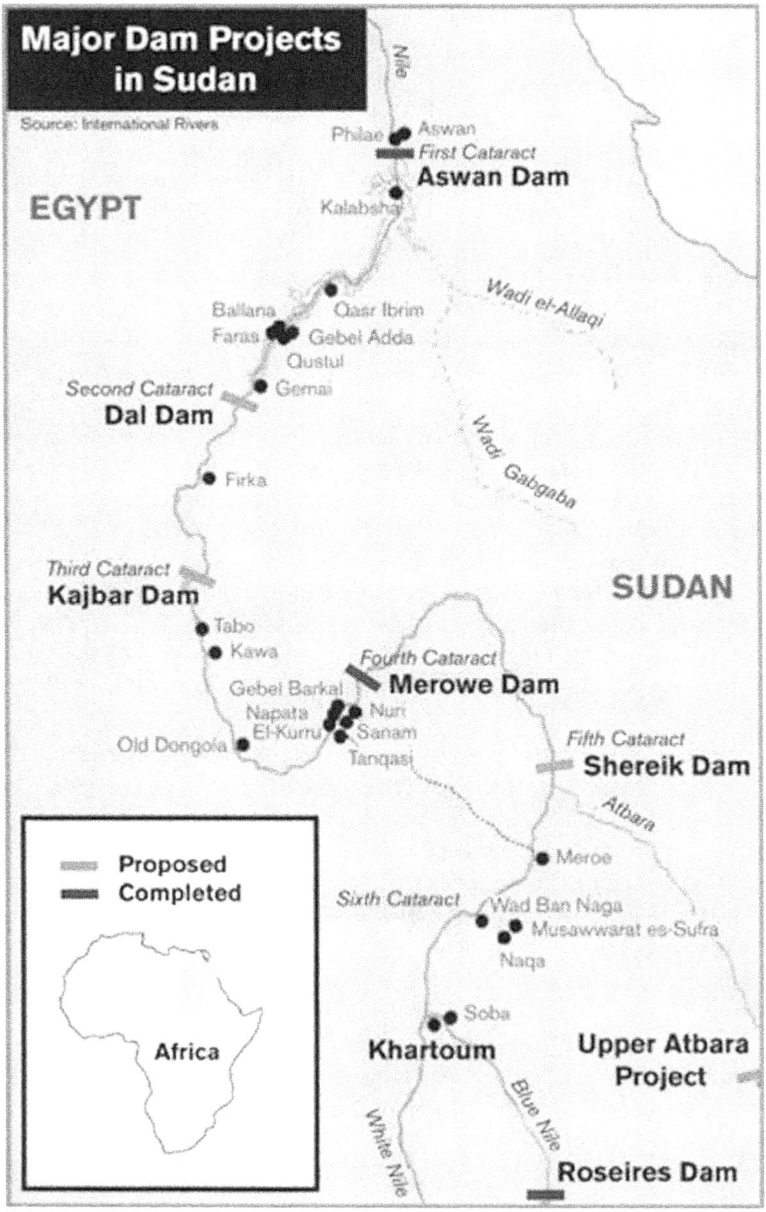

Figure 4.3 Chinese support for hydraulic projects in Sudan.

The issue of the lack of proper social and environmental assessment studies and the impact that large projects have on inhabitants can be traced back to 1899, during the construction of the Aswan Low Dam[1] when thousands of Nubians were forced to move to agriculturally unsuited areas in the West Bank of the Nile around the Aswan region. This was followed by a second and traumatic displacement of 120,000 Nubians in Egypt and Sudan in May 1964, when the second stage of the construction of the Aswan High Dam commenced. Similarly, these displaced people were taken to villages far from the Nile and away from their agricultural subsistence (Scudder, 2003, p. 1).

Today, China's lack of proper environmental studies has similarly affected the population of the basin. For instance, a proper environmental impact assessment of the Merowe and Kajbar Dam projects[2] was not carried out. As a result, many people were displaced, and a number of human rights violations occurred. On the one hand, the Merowe Reservoir, 174 km long displaced >50,000 people from the fertile Nile valley to an arid desert location (Brewer, 2008, p. 11). Similarly, the Kajbar Dam displaced 10,000 people. On the other hand, thousands of protesters were killed and some injured in protests from the affected communities (Bosshard, 2015).

While many African leaders perceive Chinese funding as an opportunity, China's lesser concerns for the environmental and social impact pose a number of threats to the populations and resources along the Nile. Bosshard (2009, p. 48), for instance, argued that Gibe III and IV[3] dams have created possible devastating effects on the environment and people living along the Omo River[4] (Bosshard, 2009, p. 48).

Since the Gibe III and IV dams are located on the same river (the Omo River), they will also negatively affect natural habitats if inappropriate environmental and social studies are conducted. The Gibe III dam and the associated decrease in water levels and seasonality of flows in the Omo River threaten the continuation of the only two options for survival in this arid environment, for example, the recessional cultivation of food along the riverbanks and livestock herding (BBC, 2010).

According to Director of the International Rivers African Programme Terri Hathaway,

> Gibe III is the most destructive dam under construction in Africa. The project would condemn half a million of the region's most vulnerable people to hunger and conflict,
>
> (BBC, 2010)

Bosshard (2009, p. 48) further argued that Gibe IV will push the fragile ecosystem of the lower Omo River Valley and Lake Turkana to the brink of collapse.

As such, the unilateral development of upstream hydropower projects through China's financial support has exerted pressure on the Nile by increasing the consumption of water; therefore, decreasing the amount available to each riparian state, and aggravating the demand-induced scarcity. This kind of scarcity could lead to conflict, as argued by Homer-Dixon's environmental scarcity theory.

Bingham (2001) posits that,

> an increase in consumption levels decreases the amount of limited natural resources available to each individual and/or state.

Homer-Dixon (1999a, p. 48) asserted that demand-induced scarcity has the potential to cause interstate conflict in that the *uncoordinated* use of a resource, such as shared water resources by one economic actor reduces its availability for others. This could lead to increased intergroup competition, which can take the form of violent conflict.

However, it is important to note that these hydropower implications are not only limited to demand-induced scarcity but also affect and exacerbate supply and structural-induced scarcities. It is important to remember that these types of scarcity are not mutually exclusive; they often occur simultaneously and interact (Homer-Dixon, 1994, pp. 8–11).

Supply-induced scarcity: Degradation and depletion of water resources in the NRB

Parts that are northeast of Africa represent one of the most water-scarce areas in the world. The NRB remains the region's only reliable source of renewable water supplies. However, the Nile's water does not flow in sufficient quantities to satisfy the future water demands of all these riparian countries (Ohlsson, 1999, p. 211).

The NRB is facing challenges of water scarcity as the NRB countries are known for their arid and semi-arid conditions. According to Ashton (2002, p. 5), the southern and the northern portions get less rainfall than their equatorial neighbouring countries. For instance, 98% of Egyptian territory is desert, with Ethiopia, Sudan, and Kenya receiving between 200 and 800 mm/year of variable rainfall (Chesire, 2010, p. 15). Karyabwite (2000, pp. 6–7) argued that rainfall in the Nile region varies from zero in the Horn of Africa to >400 mm/year in Sudan's western equatorial region. This discrepancy in rainfall negatively influences the annual flow of the Nile. The estimated decrease in water available to the Nile countries is largely due to the population increase in the basin.

Water scarcity along the Nile has intensified the potential for conflict between the upstream and downstream riparian countries. This scarcity is a result of factors, such as degradation, distribution, and population growth, among others. The effects of Nile scarcity are substantial factors that lead to internal strife and conflict between nations (Wiebe, 2001, p. 736). As the scarcity of water increases in the NRB, so does the dependency of the riparian countries on the Nile for their development (Kassa, 2010, p. 472).

The water security situation in the northeast region of Africa is worsening. The region faces challenges from climate change and the resulting continual threat of drought. For instance, between July 2011 and mid-2012, the region faced what is said to be the worst drought in 60 years. The drought caused a severe food crisis that threatened the livelihoods of approximately 9.5 million people (*The Journal*, 2012).

According to Few *et al.* (2015, p. 23), mean temperatures across the region have increased by 1°C to 3°C over the past 50 years. They further argue that evidence exists of increasing changes in extreme temperature events, such as extreme maximum temperature, warm days, warm nights, and the duration of warm spells as a result of climate change. These regional trends are supported by climate research at the national level (Few *et al.*, 2015, p. 23). Furthermore, the already precarious water situation is threatened by pollution from sprawling riverbank communities. This is the biggest problem with transboundary basins, where pollution from one area flows into another (Di Nunzio, 2013, p. 2).

The high dependence of all 11 riparian countries on the NRB, and the rapid population growth, deplete the region's already scarce water supplies as demands from agriculture, industry, and domestic use rise (Di Nunzio, 2013, p. 2). For instance, population size affects agricultural productivity. Agricultural activity is affected by resource degradation, which is affected by the land-use decisions people make based on population size (Baecher *et al.*, 2000, p. 11). Inappropriate land-use practices and farming techniques, such as overgrazing by domestic livestock, agricultural practices in irrigated and large-scale farms in rain-fed areas, and poor cultivation practices have the greatest impact on land degradation. These practices are the most common land-use practices in the NRB.

On the sixth anniversary of Africa Environment Day, the then Chairperson of the African Union (AU) commission H.E Professor Alpha Oumar Konare asserted,

> ...for us to circumvent the threats posed by environmental degradation, it is imperative that the States and Governments of Africa formulate and implement appropriate policy and legislative frameworks that would ensure the sustainable management of the environment. It is therefore essential that environmental issues are mainstreamed into the national policies planning processes in a way that would ensure maximum involvement of the local populations
>
> (AU, 2008, p. 4)

Communities concentrated along the NRB depend heavily on farming; however, the ensuing drought, famine, population growth, and land degradation have affected the NRB water resources. This is due to excessive burning for land cultivation along many parts of the Nile River, which has virtually eroded the land, making cultivation and water conservation extremely tough (Rahman, 2011). Kofi Anan argued,

> unsustainable practices are woven deeply in to the fabric of modern life. Land degradation threatens food security. Forest destruction threatens

> biodiversity. Water pollution threatens public health and fierce competition for fresh water may well become a source of conflict and wars in the future
> (UN Press Release, 2001)

Factors, such as land degradation, agriculture, industry, and domestic use are paramount to the depletion and degradation of the Nile. First, land degradation is inextricably linked to the quality, volumes and timing of water flow in the basin (Baecher *et al.*, 2000, p. 73). Land degradation has become a huge challenge within and between the Nile countries (Di Nunzio, 2013, p. 6). For instance, the Rwandan genocide led to massive displacements that left 60% of its forest damaged and 70% of its land severely degraded (Bigagaza *et al.*, 2002, pp. 51–52; Conniff, 2017). Burundi on the other hand has lost 30% of its land to degradation (Kigomo, 2003).

In 2002, 30% of Kenya's land faced severe degradation. By 2008, one-third of its population was dependent on this degraded land. This is similar to Sudan, where 76% of its population resides in degraded areas. The highlands of Ethiopia, Uganda, and Tanzania are also subject to degradation. Egypt's north-western delta faces the highest degradation due to contamination and increased salinity. This could be, in many cases, because Egypt lies downstream and receives water flow after it has been polluted upstream (Desta, 2012, pp. 10–13). Practices such as poor cultivation, especially evident in parts of the Ethiopian Highlands and the Egyptian delta, are severe and have contributed greatly to soil erosion.

Second, agricultural practices have the potential to contribute to nonpoint pollution from the use of chemical fertilisers and pesticides (Somboek and Sene, 1993; Baecher *et al.*, 2000, p. 73). Agriculture is the largest consumer of water in Egypt using about 85% of the available water. Drainage water from agricultural fields contains pollutants, such as pesticide residues, toxic organic and inorganic pollutants, salts, and treated and untreated domestic wastewater (Water Policy Programme, 2002, p. 6). About 80% of renewable water resources in the NRB are already used for agricultural purposes (UN, 2011, p. 61).

In total, >90 agricultural drains are discharging into the Nile which also includes industrial wastewater. This drainage has led to high salinisation (a build-up of salts in the soil) and saline intrusion (when the ground is saturated with salt water) in the delta. This build-up of salts in the soil means the soil cannot retain water, which prevents anything from growing (Nile Bank Initiative (NBI), 2005, p. 24). Such effects on the soil, which then cause a decline in precipitation, tend to favour soil erosion (Libiszewski, 1992, p. 3).

Given that domestic agricultural demand to sustain a growing population has exacerbated the scarcity of water, this growing demand is not only limited to the basin countries. For instance, India and Saudi Arabia have turned to large-scale land and agricultural investments in upstream countries (e.g. Ethiopia, South Sudan, and other East African countries) to help feed their growing populations (Lamere, 2012). A 2011 UN report on the use and management of the Nile, which also includes the results of a decade-long project led by the Food and Agriculture Organisation of the UN (FAO), suggests that new methods to boost agricultural productivity are required and that governments must take action to keep population growth and resource degradation from intensifying poverty (UN, 2011, p. 103).

Third, mining, although not widespread in the basin, is another source of erosion and runoff pollution. It has been noted that mines in Kilembe, Uganda, release large quantities of copper and cobalt into streams and rivers in the surrounding lakes (Kwetegyeka *et al.*, 2014, p. 482). There are also 36 industries that discharge their pollutants directly into the Nile, and into the irrigation canals. These types of industries include chemical, electrical, engineering, fertiliser, food, metal, mining, oil and soap, pulp and paper, refractory, textiles, and wood (NBI, 2005, p. 33). Fourth, in villages, where the only available water is from irrigation canals, women use the water for domestic purposes and dump the used water back into these drainages.

Egypt is a major contributor to the environmental degradation of the Nile. However, Nile water degradation in Cairo is subject to internal and external factors. Internally, Cairo's contribution to degradation is a result of rapid urbanisation, inefficient irrigation, pollution, and regional upheavals (Dakkak, 2013; Ayad, 2013). According to Dakkak (2013), the rapid population growth in Egypt, which is projected to grow from its current total of 80 million to 98.7 million by 2025, poses a threat to Egypt's water supply. This increase in population growth will require more water for domestic consumption and irrigation purposes.

Another leading factor in Egypt's water degradation is inefficient irrigation (excessive water use). This is due to half of Egypt being desert; only 6% of the country is arable agricultural land. Egypt receives <80 mm of rainfall a year, forcing the country to depend on inefficient methods of irrigation, such as excessive watering and the use of wasteful irrigation (e.g., where gallons of water are pumped over the crops).

Finally, pollution in Egypt poses an extreme threat to the water supply. For example, agricultural runoffs, industrial effluents, and municipal sewage are being recklessly dumped into the Nile, leaving its water unfit for human consumption. Agricultural runoffs contain pollutants, such as pesticides and herbicides, which negatively affect the river and the people using it. Industrial effluents containing heavy metals are highly toxic (Dakkak, 2013).

Externally, upstream of Cairo (to the south of the city), the Nile receives large flows of mostly untreated domestic, agricultural, and industrial wastewater. For instance, between the HAD and Cairo, 43 towns with a population exceeding 20 million and approximately 2,500 villages discharge their waste into the river. In addition, most of the residents in the region depend on irrigated agriculture for their livelihood, and 29 billion cubic metres (BCM) of drainage water loaded with fertilisers, pesticides, and organic material is returned to the Nile upstream from Cairo annually. More than 50 major factories discharge approximately 250 BCM/year of industrial wastewater (Ayad, 2013).

Without a cooperative framework, countries along the Nile such as Burundi and Ethiopia are taking advantage of the Arab spring and the political strife that has engulfed Egypt and is gaining more control over the Nile. With the Nile supplying 95% of Egypt's freshwater, losing some of the water supply can cause additional problems for Egypt (Dakkak, 2013).

Environmental degradation and depletion in Cairo pose an alarming threat of conflict in the region. In the near future, high levels of water degradation assume that Cairo will consume approximately 20% more water than it already does. This supposes that other riparian countries will be left with little, if any, water to consume for developmental purposes. This minimal consumption by other riparian countries at the hands of their downstream neighbour, which has maintained dominance over the use of the Nile for decades, makes the potential for disputes and conflict more likely. The water issue in Egypt is rapidly assuming alarming proportions. With the lack of effective and efficient water conservation techniques, other riparian countries' water supplies are being highly affected (Dakkak, 2013).

Another cause of NRB depletion is global greenhouse emissions. This argument is consistent with the effects of climate change in Ethiopia. Climate change in Ethiopia has led to both a more variable climate and more extreme weather events such as the drought that engulfed the entire country between 1970–1980 (Rahman, 2011). Kim and Kalaurachchi (2009, p. 1363) argued that global warming because of climate change and greenhouse emissions was one of the leading contributors to the current decline in the NRB water resource. This severe impact of global warming is also significant to the reduction of underground water in the Blue Nile in Egypt and Sudan.

Development along the Nile has led to water resource pollution by many riparian countries and dates back to the Ethiopian and Eritrean wars in the late 1990s. Military accoutrements and missile deposits were thrown into the Nile River, polluting a substantial part of the river basin (Rahman, 2011).

The way the land, water, and watershed have been degraded in the NRB has produced and sustained poverty. These resources fail to provide a sustainable or desired flow of water to neighbouring communities and countries, which exerts even greater pressure on the available water resources for agricultural and industrial purposes. Another problem related to degraded watersheds is that they often lead to erosion and therefore export water of a poor quality that affects downstream countries (Baecher *et al.*, 2000, p. 39).

The environmental implications for the Nile are controversial as far as China's involvement in the basin is concerned. This raises the question of whether China is addressing the overwhelming pressures on water supplies in the NRB or increasing them. These unilateral projects in upstream countries are negatively impacting the sufficient transfer of water, which is detrimental to ensuring peace and security in the region.

Soil conservation in the basin has been heavily affected because effective environmental and social impact assessments were not carried out. For instance, the displacement of people during the construction of the Merowe project hugely affected agricultural activity. A concern when it comes to the retraction of arable land and displacement is that agricultural activities cannot be restored at the new resettlement sites because the soil is too poor and does not compensate for a realistic source of income despite the provision of irrigation water (Brewer, 2008, p. 11).

In 2006, the Swiss Federal Institute of Aquatic Science and Technology, in an independent review of the Merowe project, outlined that strongly fluctuating water levels and sedimentation would have a serious negative impact on aquatic ecology, water quality and public health. As a result, Europe, Canada, and other financial institutions withdrew funding for this project. However, the project went ahead because of Chinese funding (Brewer, 2008, p. 11).

In the entire NRB, 1 million ha of fertile land has been destroyed because of hydropower stations and dams. The loss of arable land was also witnessed in Ethiopia through massive landslides that took place on the Tekeze River[5] because of the construction of the Tekeze hydroelectric dam. Many environmentalists, including Richard Leakey, have expressed concerns about the possible impact of the Omo River dams on Lake Turkana[6] (Greste, 2009).

According to an International Rivers environmental group report, the completion of the Gibe III dam will negatively impact the water quantities for communities that live near Lake Turkana in neighbouring Kenya. The survey predicts that in these areas, including areas near the lake, there will be more conflict if the water levels decline (GeeskaAfrika Online, 2015). This prediction is not limited to riparian countries but also applies to riparian countries. There is currently no comprehensive integrated water resources management plan for the Blue Nile Basin and no adequate monitoring infrastructure or erosion-prevention efforts (Veilleux, 2013, p. 6).

According to Ugandan environmentalist Muramuzi (2011), Lake Victoria,[7] the largest lake in Africa and the most renowned source of the Nile, is slowly and gradually shrinking due to the increasing number of hydropower projects and irrigations in Uganda, which draw large amounts of water out of the lake.

With millions of people in the basin being severely affected by soil degradation, these uncoordinated development projects will further affect water security, soil conservation and food production to supply a growing population. The soil degradation has rendered large amounts of arable land unusable for food production (Muramuzi, 2011).

It is evident that these unilateral projects have produced a supply-induced scarcity in that they have intensified the degradation and depletion of the Nile. According to Homer-Dixon's environmental scarcity theory, supply-induced scarcity in the Nile is an increasingly salient element of interstate tensions and under unfavourable conditions will lead to violent conflicts (Homer-Dixon, 1999a, p. 48). Muramuzi (2011) further asserted that these unilateral developments, although a positive achievement for upstream countries' developments are extracting more water than the Nile can provide. The projects will exert pressure on already exposed water beds downstream (as evidenced in the Nile's cataracts) and upstream. These uncoordinated developments will therefore result in lower water tables and rivers running dry due to unsustainable use (Van der Molen and Hildering, 2005, p. 135).

Uneven distribution of water resources in the NRB

International river basins have become tense due to competitive exploitation. Consequently, disputes have become inevitable (Kliot *et al.*, 2001, p. 230). However, the argument that water scarcity will lead to conflict in the NRB is contested by an opposing view that water scarcity is likely to force cooperation between riparian countries. This would be true only if the underlying cause of water scarcity in the NRB was not the politics of distribution (Pakes, 2013, p. 434).

One of the most contentious issues related to the Nile River is the use of available water resources (UNEP, 2006, pp. 2–4). Water scarcity is not only a result of the amount or scale of rainfall but also of the agreements that govern the Nile. These agreements have rendered the use of water between the upstream and downstream riparian countries unequal.

The use of the Nile has always been and remains a controversial issue. Water is already scarce in some parts of the basin. The current distribution of water in the area has serious global political consequences and is fundamental to the developmental processes of the respective countries (Oestigaard, 2012, p. 25). This uneven distribution also bears many local political, economic, and social implications.

According to Haftendorn (1999, p. 10), the impact of resource distribution, specifically water, carries many political implications in local, state, regional, and international spheres. In shared basins, the lower or upper-lying riparian country can control the quantity and quality of water flow using military might and hydrological situation. Political actors may often also use resources as objects of military and political action to extend their influence in such regions (Conca *et al.*, 2000, p. 1).

An example of this is the defining influence of Egyptian foreign policy on states in the basin. Egypt, being the stronger party, provokes, initiates, or prevails with military action against its weaker adversaries, thereby increasing the potential for conflict. The issue of controlling the Nile has been central to political decision-making in Cairo (Hassan and Rasheedy, 2011, p. 149).

The Nile, as an internationally shared river that extends across political borders, has created political, social, and economic tensions and disputes between upstream and downstream countries concerning the distribution and use of water. These include misunderstandings or lack of beneficial agreements about the allocation of water (Conca *et al.*, 2000, p. 1).

Egypt's close relationships with a succession of major powers also helped it to exert its influence over the Nile. The UK, until the 1950s, the Soviet Union until the mid-1970s and, currently the US, have all provided political and financial support to cement Egypt's position in the basin, through legal treaties if possible. They have also simultaneously influenced the construction of infrastructure for power generation, storage and irrigation (Verhoeven, 2013).

The Nile, therefore, became subject to numerous agreements. Many bilateral treaties were concluded between Egypt, the UK, and other foreign or European powers to regulate the use of the Nile's water (Okoth-Owiro, 2004, p. 1). Egypt became important to the UK for the following reasons.

- Egypt, through its control of the Suez Canal, became the easiest and quickest way to India as opposed to the Cape of Good Hope.
- Egypt is strategically located between the UK and India. This strategic position became important for the UK to maintain communication with India.
- During the American civil war, the UK shifted its attention to Egypt for its cotton.
- The Nile River was presented as a source of water and flood control in the move from the traditional flood-fed methods of irrigation to an increased reliance upon perennial irrigation (Luscombe, 2012). In this instance, the UK, which had become reliant on the agricultural exports of the Nile region, saw the allocation between Egypt, British East Africa, and Sudan as the means for irrigation (Knobelsdorf, 2006, p. 5).

Godana (1985, p. 176) also argued that to regulate European colonial rule over most of the Nile region, it became necessary to establish treaties and instruments that would attach the water rights and obligations within the basin to Europe. However, it is important to consider prior treaties regarding the use of the Nile, which culminated in the current nature and context of the 1929 and 1959 agreements.

A series of agreements deemed responsible for structurally denying upstream riparian countries access to, and use of, the Nile include:

- The 1891 Protocol between the UK and Italy to demarcate their respective spheres of influence in Eastern Africa.
- The 1902 Anglo–Ethiopian Treaty to delimit the frontier between Ethiopia and Sudan.
- The 1906 Treaty between the UK and the Independent State of Congo to redefine their respective spheres of influence in eastern and central Africa.
- The 1906 Tripartite Agreement and set of declarations between the UK, France, and Italy.
- The 1925 Exchange of Notes between Italy and the UK.
- The 1929 Agreements: exchange of note between His Majesty's government in the UK and the Egyptian government on the use of the Nile for irrigation.
- The 1959 Agreement: the agreement between the Republic of Sudan and the United Arab Republic for the Full Utilisation of the Nile.

This series of agreements shared the principle of protecting Egyptian interests in the Nile region. They stipulated that countries would not construct or cause to be constructed irrigation or any work that might negatively affect the flow of the Nile to Egypt. The UK, France, and Italy argued that they were acting on behalf of upstream riparian countries. For instance, with the fear that the Italians, acting on behalf of Eritrea, may construct on the upper Nile, the UK warned Italy to avoid any access or construction on the Nile in exchange for the recognition of an Italian sphere of influence in the Ethiopian Highlands (Yacob, 2014).

Article 3 of the 1891 protocol sought to protect Egyptian interest in the Nile's Atbara River, the upper reaches of which were within the newly acquired Italian possession of Eritrea (Okoth-Owiro, 2004, p. 6). Provisions of the article stated:

> The Italian Government engages not to construct on the Atbara, in view of irrigation, or any work which sensibly modify its flow into the Nile,
>
> (Degefu, 2003, p. 95)

It was through this demarcation process that the UK, to assure the Italians' acquiescence, agreed to recognise Ethiopia as an Italian sphere of influence, thus formally denying Ethiopia its independence (Degefu, 2003, p. 95).

Article 3 of the 1902 Anglo–Ethiopian Treaty stated:

> His Majesty the Emperor, Menelik II, King of Kings of Ethiopia, engages himself towards the Government of His Britannic Majesty not to construct or allow to be constructed, any works across the Blue Nile, Lake Tana or the Sobat, which would arrest the flow of their waters into the Nile except in agreement with his Britannic Majesty's Government and the Government of the Sudan,
>
> (Rubenson, 1991, p. 302; Okoth-Owiro, 2004, p. 7; Yihdego, 2013, p. 1)

Article 3 of the 1906 Treaty between the UK and the Independent State of Congo stipulated:

> The Government of the Independent State of Congo undertakes not to construct or allow to be constructed any work over or near the Semliki or Isango Rivers, which would diminish the volume of water entering Lake Albert, except in agreement with the Sudanese Government,
>
> (Treaty Series no.4, 1906:c36; Okoth-Owiro, 2004, p. 7)

The 1906 Tripartite Agreement, on the other hand, acted as a confirmation of the terms of the protocol between the UK and Italy, and the Anglo–Ethiopian treaty. The three parties recognised the principle of non-interference with the flow of the Blue Nile. Article 4 (a) of the tripartite treaty stated:

> France, Great Britain and Italy shall make every effort to preserve the integrity of Ethiopia, in an effort to safe guard the interests of Great Britain and Egypt in the Nile Basin, more especially as regards the regulation of the waters of that river and its tributaries,
>
> (Yihdego, 2013, p. 1)

Under the 1925 Exchange of Notes between Italy and the UK, Italy recognised the rights of Egypt and Sudan in the Blue Nile, the White Nile, and their tributaries.

It agreed that it would not engage in any construction that might modify these rivers' flow into the main river (Okoth-Owiro, 2004, p. 7). The UK's interest was not limited to the river alone, but included the control of the source, for instance, Lake Tana in Ethiopia, which is the largest lake. Italy stated:

> … we have noted Britain's determination to control the Blue Nile for the Benefit of Egypt and the Sudan,
>
> (Degefu, 2003, p. 111)

According to Knobelsdorf (2006, p. 5), this series of negotiations and understandings became what is known as the 1929 Nile Water Agreements.

> The 1929 Agreements: Exchange of note between His Majesty's government in the United Kingdom and the Egyptian government on the use of Waters of the Nile for irrigation.

The sole purpose of the 1929 Nile Agreements was to guarantee the 1920 estimates made by the Nile Project Commission, with representatives from India, the UK, and the US (Lumumba, 2007, p. 22). With the Nile Rivers' average flow of 84 billion BCM/yr, these estimates suggested that Egypt would need 48 BCM/yr and that Sudan's irrigation would survive on the Blue Nile alone, which accounted for approximately 4 BCM/yr. This left 32 BCM/yr unallocated (Wolf and Newton, 2008). Due to the fluctuation in the river, it was agreed that during any change in the flow, gain or shortfall, the Nile's water would be divided evenly between Egypt and Sudan (McCaffrey, 2003, p. 265; Okoth-Owiro, 2004, p. 8; Lumumba, 2007, pp. 12–13).

These estimates led to the Nile Waters Agreements on the allocation of water between Egypt and Sudan, also known as the Anglo–Egyptian Treaty. The 1929 Nile Agreement was signed on 7 May between Egypt and the UK (the then colonial master of Sudan) to share the Nile's water. These agreements were signed as a means for both countries to engage in comprehensive large-scale control of the Nile (Kieyah, 2007, pp. 7–8).

According to Dunoff and Trachtmann (1999, pp. 24), the agreements granted Egypt extensive use of the Nile. They stipulated that Egypt has a right to use the Nile and a property rule to protect this right. This gave Egypt exclusive propriety rights to the Nile without obligation, consent, or even voluntary transfer of property rights from Egypt to other riparian countries (Kieyah, 2007, p. 19).

Through these agreements, Egypt was guaranteed a claim to the entire timely flow (from January to July); rights to on-site inspectors at the Sennar Dam built in Sudan; and that no works would be developed along the river or on any of its territories that would threaten Egyptian interests (Wolf and Newton, 2008). These agreements also granted Egypt power over any upstream countries' projects in the southern Nile that concerned the Nile, and Sudan was bound by Egypt's approval before undertaking any irrigation project.

To further Egypt's hegemony over the Nile, Note 1, par. 4(b) of the agreements also required Sudan or countries under the British administration to obtain the consent of the Egyptian government before undertaking any irrigation work or installing electric generators. This part of the agreement concerned the quantity of water flowing into Egypt from the branches and lakes of the Nile (International Water Law Project, 2014a). It was another clear symbol of the UK's lack of representation and the compromise of its upstream colonies. According to Oloo (2007, p. 101), in the process, Egypt carried out a series of major water projects that enabled it to acquire extensive control over the Nile by appropriating large portions of the Nile's water and bringing the flow within its sovereign jurisdiction.

The enactment of the 1929 Nile Agreements enabled Egypt to use Note 2, par. 4 of the agreements to argue the following based on natural and historical rights (Knobelsdorf, 2006, p. 6):

> Her Majesty's Government in the UK has already recognized the natural and historical right of Egypt to the waters of the Nile....Her Majesty's Government in the United Kingdom considers the observance of these rights as a fundamental principle of the policy of Great Britain...
>
> (International Water Law Project, 2014a)

Although the 1929 Nile Agreements served as the basis for the principles of Nile water allocation, they did not guarantee the full use or exploitation of the Nile's water; hence the need for a revised agreement. The 1929 Nile Agreements were revised on 8 November 1959 through the 1959 Nile Agreement.

> The 1959 Agreement: Agreement between the Republic of the Sudan and the United Arab Republic for the Full Utilization of the Nile Waters.

In 1959, Egypt and Sudan concluded an agreement. This bilateral agreement was entirely exclusive in that it allocated the entire flow of the Nile's water at Aswan to Egypt and Sudan; therefore, effectively excluding other Nile riparian states (Knobelsdorf, 2006, p.8; Lino, 2013, p. 12).

Abseno (2013, p. 146) argued that the Nile is recognised by both countries as their respective historic and acquired rights. According to Caponera (1993, p. 661), a nation enjoys the right to water that has been currently or historically used by that nation. Acquired rights according to Sinclair (1996, p. 87), means a right acquired under the law to impound, divert, or use water.

Contrary to the 1929 Nile Agreements, the 1959 Nile Agreement referred to other riparian countries. Article 5, par. 2(a) states:

> As the riparian states, other than the two Republics, claim a share in the Nile waters, the two Republics have agreed that they shall jointly consider and reach one unified view regarding the said claims,
>
> (International Water Law Project, 2014b)

Ethiopia, for instance, reserved its rights to the Nile in 1956, 1977 and 1980. Kenya, Uganda and Tanzania also reserved their rights in the *Nyerere Doctrine of Treaty Succession*, refusing to be bound by colonial-era agreements (Kliot, 1994, p. 86). However, the agreed *unified view* of Egypt and Sudan was to reject all these claims.

The adoption of the 1959 Nile Agreement replaced allocations that were set forth in the 1929 Nile Agreements and established a *joint* bilateral full use of the Nile's water between Egypt and Sudan (Demeke 2013). The two purposes mentioned in the introductory stipulations of the 1959 Nile Agreement specify Egypt and Sudan's intentions to claim full control over the Nile's water and to make solid future claims on the Nile.

According to these introductory stipulations:

> The Nile River projects require for their execution and administration, full agreement and cooperation between the two Republics in order to regulate their benefits and utilize the Nile waters in a manner which secures the present and future requirements of the two countries; and, as the Nile waters Agreement concluded in 1929 provided only for the partial use of the Nile waters and did not extend to include a complete control of the River waters, the two Republics have agreed on the complete control of the River waters,
>
> (International Water Law Project, 2014b)

Nonetheless, the 1929 Nile Agreements remained the yardstick for the 1959 Nile Agreement. As such, the 1959 Nile Agreement did not differ much from the 1929 Nile Agreements and continues to render full control over the Nile to Egypt, with huge economic benefits and the bonus of hydropower (Allan, 1999, pp. 4–5).

In short, both agreements give Egypt and Sudan absolute rights to use 90% of the river's water, leaving the upstream countries to share 10%, which created an unsustainable solution. This solution impairs developments in upstream countries, which are forced to abandon certain projects due to limited access to water or to get approval from Egypt to use the Nile's water or construct water-related projects (Kameri-Mbote, 2007, p. 3).

The 1929 Nile Agreements granted most of the Nile's water to Egypt, with 48 BCM/yr of the 84 BCM/yr total average of the Nile's water flow, and granted Sudan 4 BCM/yr. However, article 1, par. 1–2, and article 2, par. 3–4 of the 1959 Nile Agreement ensured that the unallocated 32 BCM/yr was also divided between Egypt and Sudan according to their respective needs (Wolf and Newton, 2008; International Water Law Project, 2014b).

According to these provisions, the average flow of the river is 84 BCM/yr. The agreement considered evaporation and seepage to account for 10 BCM/yr, leaving 74 BCM/yr to be divided between Egypt and Sudan. The agreement argued that Egypt's *acquired rights* would account for approximately 48 BCM/yr

Table 4.1 Water allocation from Nile agreements

Nile average flow = 84 BCM/year	*1929 Nile Agreements (BCM/year)*	*1959 Nile Agreement (BCM/year)*
Egypt	48	55.5 (48 + 7.5)
Sudan	4	18.5 (4 + 14.5)
	Leaving 32 BCM/year unallocated	Leaving 10 BCM/year for evaporation and seepage

Source: Wolf, 1996, p. 3.

and Sudan would acquire 4 BCM/yr. The remaining benefits of approximately 22 BCM/yr are divided by a ratio of 7½ BCM/yr for Egypt and 14½ BCM/yr for Sudan (International Water Law Project, 2014b) (Table 4.1).

As indicated in the agreement, these calculations are based on historical rights and acquired rights. It is difficult to determine, monitor, and quantify the Nile River flow or its actual total annual flow, since it varies by season and year The average annual BCM used to argue the allocation in the agreement took expected seasonal fluctuations into account (Knobelsdorf, 2006, pp. 6–8).

Irrespective of the burden laid upon upstream countries to obtain Egypt's approval if they desired to develop projects, the 1959 Nile Agreement also made the process complex by requiring upstream countries to obtain approval not only from Egypt but also from a technical committee based on technical oversight and contractual supervision. Article 5, par. 1 and 2 of the 1959 Nile Agreement stipulate:

> If it becomes necessary to hold any negotiations concerning the Nile, with any riparian state, outside the boundaries of the two Republics, the Governments of the Sudan Republic and the United Arab Republic shall agree on a unified view after the subject is studied by the said Technical Commission,
>
> (International Water Law Project, 2014b)

The said unified view:

> shall be the basis of any negotiations by the Commission with the said states; and, If the negotiations result in an agreement to construct any works on the river, outside the boundaries of the two Republics, the joint Technical Commission shall after consulting the authorities in the Governments of the States concerned, draw all the technical execution details and the working and maintenance arrangements. And the Commission shall, after the sanction of the same by the Governments concerned, supervise the carrying out of the said technical agreements,
>
> (International Water Law Project, 2014b)

According to Katz (2013, p. 1256), although the 1959 Nile Agreement replaced, altered, or updated the allocations in the 1929 Nile Agreements, it did not entirely replace the obligations of the 1929 Nile Agreements. The updated allocations were merely a response to the political and agricultural changes and demands since the signing of the 1929 Nile Agreements. Like the 1929 Nile Agreements, the legal consequences of the 1959 Nile Agreement are totally unacceptable (Demeke, 2013). For instance, following the 1959 Nile Agreement, Egypt's plans included substantially increasing its use of Nile water. This was to overcome the challenges of increasing population growth and growing pressure over *old* lands in the Nile valley and delta (Cascao, 2009, p. 249).

These challenges were addressed by expanding three major agricultural projects and a government policy of moving people out of these old lands into the newly reclaimed lands in the desert where these projects were taking place (converting these deserts into land for agricultural and industrial development). These projects have been ongoing horizontally since the late 1990s and include the West Delta Irrigation Project, the North Sinai Agriculture Development Project, and the South Valley/Toshka Development Project, all of which aim to reclaim thousands of ha of land (Cascao, 2009, p. 249).

Of significance is the South Valley/Toshka Project, which started in 1997 and has since attracted a high level of criticism from upstream riparian countries. This is mainly because the project aimed to reclaim 1.5 million acres of land estimated to require 5 BCM/yr of water. For upstream countries, this means that Egypt will use more Nile water than it is already using and attain legal rights to additional water allocations (Schilling, 2013).

Egypt's use of natural and historic rights has constrained relations between upstream and downstream countries since these rights are contradictory. Al-Din Amer (1997, p. 381) expounded on these rights, arguing that the principle of prior appropriation or the international law principle of *absolute territorial integrity* (acquired rights) argues to *protect the rights of use for the country that puts the water into use first*. Therefore, downstream countries are given the power to accuse upstream countries of illegally taking measures that are disadvantageous to their interests. However, the principle of *absolute territorial sovereignty* (natural rights) regards water bodies as an integral part of the state's national territory; therefore, protecting the sovereignty of countries within which watersheds, lakes and rivers originate.

Egypt has relied heavily on its military might to safeguard its position on these agreements and its rights to the Nile. For instance, in 2005 Egypt's military expenditure was more than twice the military expenditure of the other nine riparian states together (Adar, 2011, p. 183) (South Sudan had not yet been formed). This is a clear reflection of how central the security of the Nile is to Egyptian military plans and Egypt as a whole. According to Yacob (2004, p. 2), the Egyptian High Command even went out of its way to establish contingency plans for armed intervention in each country in the NRB since they are all perceived as national security threats to Egypt.

These military plans, also known as *Waraa-elhidoud*, include handling direct threats to the flow of the Nile's water. Such behaviour can be traced back to the early nineteenth century when Mohammed Ali was rebuilding the Egyptian army. However, today the Nasser Military Academy in East Cairo appoints full-time staff to review and adjust these military plans to changing modern circumstances (Yacob, 2004, p. 2). This is proof of Starr's (1991, p. 19) argument that,

> water security will soon rank with military security in the war rooms of defence ministries.

It is such behaviour that has forced both upstream and downstream riparian countries to a state of preparedness for war, in case there is one (Ashton, 2002, p. 1).

Egypt has, therefore, structurally deterred upstream riparian states from using the Nile by these agreements (Chesire, 2010, p. 13). This has left upstream countries in a position where they cannot develop or even get enough water to their people (Mayton, 2012).

This is despite upstream countries being major contributors to the Nile's water (Figure 4.4). Ethiopia and Eritrea for instance, face challenges of droughts, famine and other severe implications that come with this lack of access to the Nile (Yacob, 2014).

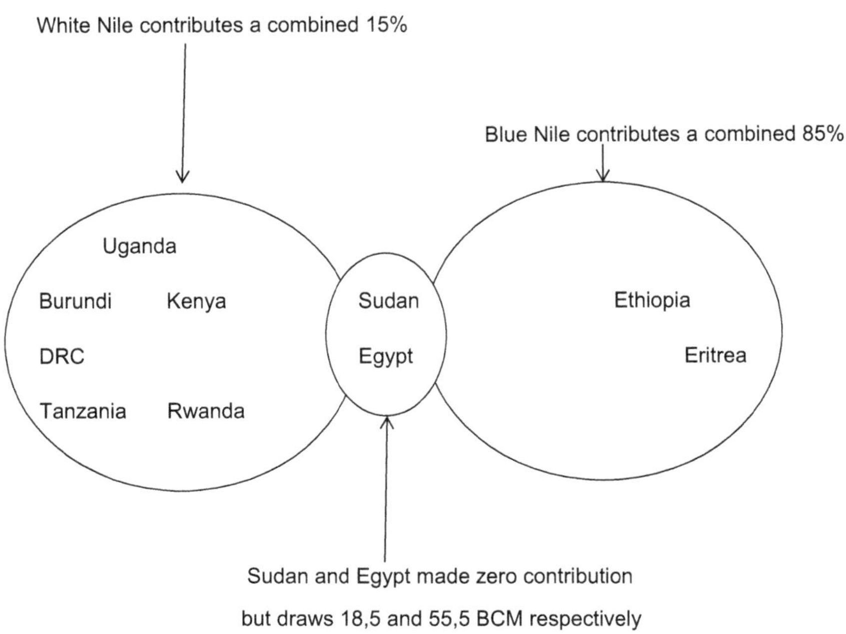

Figure 4.4 Contribution of upstream countries in the Nile.

However, any means to develop and use the Nile has resulted in bitter disputes between Egypt and upstream riparian countries. For instance, Ethiopia has pushed its demands to develop water resources to address its challenges through irrigation schemes and the construction of dams such as the Grand Ethiopian Renaissance Dam (GERD) (Rahman, 2011). The now-completed GERD, which is built right on the headwaters of the Blue Nile on the Ethiopian–Sudanese border, is intensifying insecurities for Egypt, which receives 59% of its waters from the Blue Nile (Conniff, 2017).

The anticipation was, according to Conniff (2017), that filling the GERD could take approximately between 4 and 7 years. During this fill, Egypt could experience a cut of 25% in the Nile's freshwater flow. Moreover, this is on the background of an already serious country-wide freshwater shortage in Egypt (Conniff, 2017). According to Al Jazeera (2022), the GERD Reservoir's total capacity is 74 BCM and the process of filling this reservoir began in 2020.

With the first and second phases of filling the GERD being completed on 20 July 2020 and 19 July 2021, respectively (Anna, 2020; BBC News, 2022), the reality of any nascent threat to the NRB and its inhabitants' peace, stability, and water security will soon be realised. This is because the Nile will now have two of the world's largest dams (i.e., the HAD and GERD – both located in Egypt and Ethiopia, respectively), and in actuality that as of now there is no agreement on the management and operation of both dams, and at the face of severe water scarcity (Wheeler *et al.*, 2020, p. 1).

The first filling of the GERD retained 4.9 BCM (Basheer *et al.*, 2020, p. 279), while the second filling retained 13.5 BCM of water, which approximates the current impoundment capacity of 18.4 BCM (Wheeler *et al.*, 2020, p. 7). The GERD project alone has strained relationships among Nile countries. Although analysts argue that water flows into Egypt could be reduced depending on the initial storage of the HAD and the filling of the GERD (Verhoeven, 2013; Elsayed *et al.*, 2020, p. 14); however, the precise impact of the dam on the downstream countries is not known (Wheeler *et al.*, 2016, p. 617).

According to Wheeler *et al.* (2020, p. 3), the downstream consequences of filling the GERD are difficult to assess if six key factors are not factored into the equation. These include: (1) rainfall and flow in the basin during filling; (2) how quickly Ethiopia fills the GERD Reservoir; (3) how hydropower is generated at the GERD, which equally depends on developments in the regional power grid; (4) how the GERD Reservoir filling influences Sudan's withdrawals; (5) the initial storage level in the HAD Reservoir, given the fact that it was almost full at 178.4 m above sea level on the 24 August 2020; and (6) how the HAD is operated during the filling period. However, studies show that only through close and continuous coordination, the risks of negative impacts can be minimised or eliminated.

This ultimately resulted in waves of disputes and tensions, as anticipated by the environmental scarcity theory (Homer-Dixon and Blitt, 1998, p. 6; Homer-Dixon, 1999a, p. 4). These disputes and tensions are evident between Egypt and Ethiopia.

In 1977, Egypt threatened Ethiopia with a war on numerous occasions. For instance, in the same year, Egyptian President Sadat declared:

> if Ethiopia proceeded there is no alternative but to use our force. We will retaliate when something happens but we have to be ready with plans and alternatives to firmly stop any action,
>
> (Adar 2011, pp. 182–183)

In May 1978, President Sadat warned:

> We depend upon the Nile 100% in our life, so if anyone, at any moment thinks to deprive us of our life we shall never hesitate to go to war because it is a matter of life or death,
>
> (Waterbury 1979, p. 78)

In 1979 Sadat reiterated

> the only matter that could take Egypt to war again is water.

In 1991, former Egyptian Defence Minister Mohamed Hussein Tantawi Soliman also gave an assurance of Egypt's readiness to use force, if necessary, to protect its control of the Nile.

The militancy expressed by Egypt also relates to its dominance and stature at the World Bank, as the main financer of the NBI. Egypt is seen as pushing these international donor bodies. The World Bank has played a major role in denying other riparian states access to the Nile. In 2004, the Ethiopian Minister for Trade accused Egypt of using undiplomatic strategies to control Ethiopia's development projects on the Nile. He stated,

> Egypt has been pressuring international financial institutions to desist from assisting Ethiopia in carrying out development projects in the Nile basin,
>
> (Braune and Xu, 2010, p. 232)

For instance, the World Bank opposed Ethiopia's intention to construct the GERD Dam. In 2009, during the Alexandria meeting, World Bank officials said they would not fund any projects without the approval of Egypt, stating,

> Egypt is the leading country in this consortium and the World Bank will not get behind any initiative that leaves them [Egypt] out,
>
> (Green Prophet, 2010; Mayton, 2012)

This was not the first time that Egypt had influenced financial institutions to desist from assisting Ethiopia. In 1990, Egypt blocked an African Development Bank loan to Ethiopia for new irrigation projects (Moffet, 1990, p. 5).

On the other hand, Ethiopia's minister of water resources announced at a 1997 conference in Addis Ababa on the NRB Action Plan,

> as a source and major contribution of the Nile, Ethiopia has the right to have an equitable share of the Nile waters and reserves its rights to make use of its waters.

Ethiopia's foreign minister stated in 1998,

> there is no earthly force that can stop Ethiopia from benefiting from the Nile,
> (Shinn, 2006, p. 3)

In 2004, Egypt restricted Tanzania from constructing the Lake Victoria pipeline claiming that, according to the 1929 Nile Agreements, Tanzania could not construct any project without British permission. This project would have been of huge benefit to about 400,000 of Tanzania's north-western citizens. Egypt threatened to bomb the construction site arguing that it needed the water to flow northward into the Aswan Dam (Di Nunzio, 2013, p. 6).

Although the nature of the 1959 Nile Agreement supposed a cooperative relationship between Egypt and Sudan, the latter also received its fair share of military threats from Egypt (Deng, 201, p. 39). In June 1995, President Mubarak announced,

> Those who play with fire in Khartoum … will push us to confrontation, [to] defend our rights and our lives,
> (Swain, 2011, p. 691)

Deng (2011, pp. 52–53) further argued that Sudan received more threats from Egypt than any other riparian country; hence, it is also subject to Egyptian manipulation and plots. This is highlighted by Egypt's interference in aiding the rebel Sudanese People's Liberation Army in southern Sudan (Klare, 2002, p. 154).

Irrespective of these threats, it is evident that Ethiopia will stop at nothing to push for its share of the Nile. It is also evident that Ethiopia has lost confidence in the "Do-no-harm" agreement signed in 2015 between Egypt, Ethiopia, and Sudan due to the agreement's lack of providing a formal agreement on how to share the Nile. The absence of a formal agreement regarding the actual share of the Nile, particularly by Ethiopia, has appeared to only serve Egypt's interests (Conniff, 2017).

It can, therefore, be concluded that these agreements have systematically denied upstream riparian countries access to the Nile by imposing obligations on their colonial administrations not to construct the Nile. For instance, Italy, acting on behalf of Eritrea and recognised as a sphere of influence in Ethiopia, agreed to these obligations. Belgium, acting on behalf of Burundi, the DRC and Rwanda, and the UK acting on behalf of Kenya, Tanzania, and Uganda, but excluding

Sudan, also agreed to these obligations. This rendered Egypt control over the Nile while compromising these upstream riparian countries' access to the Nile.

Today, all these upstream countries are subject to the same obligations and, because of their lack of access to the Nile or the obligation not to construct the Nile, are faced with serious humanitarian and developmental crises. The UK's representation in all the previous negotiations was meant to include the concerns of its colonies and mobilise upstream development (Paisley and Henshaw, 2013, p. 5).

However, the upstream riparian countries reject this claim as these agreements were imposed on them and the populace had little input in designing or approving the agreements. The agreements served primarily Egyptian and a minor British–Sudan interest, thus binding the UK-controlled upstream riparian countries to Egyptian interests (Knobelsdorf, 2006, pp. 1–5).

In addition, treaties before the 1929 and 1959 agreements have played a huge role in both these agreements, such as *securing recognition of the principle that no upper-basin state had the right to interfere with the flow of the Nile in respect of Egypt*, (Okoth-Owiro, 2004, p. 7). These agreements have been a source of structural-induced scarcities in the NRB.

The agreements outlined in this chapter provide an indication of the structural-induced scarcity as argued by Homer-Dixon's environmental scarcity theory. This was done to provide the theoretical linkage between structural-induced scarcity and the potential for conflict in the NRB. This linkage is argued on the basis that property rights or other institutions are established to prevent access to water by other actors. The scarcity and grievances that may result can reinforce existing stress factors and play a role in inciting tensions and disputes (Homer-Dixon, 1999a, p. 48; UNEP, 2012, p. 14). Tvedt (2004, p. 480) argued,

> these water sharing treaties have kept Nile Basin countries into conflict over the regions' most precious resource.

China's involvement in the hydropolitical development of the Nile has also exacerbated structural-induced scarcity in the basin. According to Homer-Dixon (1999a, p. 48), structural-induced scarcity is a major cause of violent conflicts in that it deprives certain water users of access to water. Unilateral hydropower developments in upstream countries have also become a major source of structural-induced scarcity and resulted in an unequal distribution of the Nile. Given that upstream countries are gradually becoming more powerful water users of the Nile, it is evident that soon they will confiscate a larger part of the Nile.

Although these unilateral developments have bypassed the colonial agreements that governed the Nile, they did not address structural-induced scarcity in a mutual sense; rather, they exacerbate it. Unilateral hydropower developments have increased the unequal distribution of the Nile. Because of these developments, Egypt will experience physical water scarcity (Martens, 2011, p. 9).

The basis of the threats posed by unilateral hydropower developments upstream is that these projects severely affect the water allocation formula on which Egypt is heavily dependent, and has been for decades (Swain and Jamali, 2011, p. 9). As argued before, Egypt's dominance of the Nile was detrimental to the peace and security of the region; however, so are these unilateral developments. This chapter argues that an equitable allocation is only possible through a negotiated treaty.

Because Egyptian interests were not considered when constructing these projects, environmental studies to determine the impact of these developments on Egypt's water security were not conducted (Hegazi, 2011, pp. 33–34). As a result, the issue of uneven distribution will continue to grow rapidly. For instance, the hydropower station at Karuma Falls threatens the natural flow of the Nile (Muramuzi, 2011).

This may be the beginning of a new phase of competition over the use of the basin's transboundary water resources. The following section will determine the existence and effectiveness of regimes in the Nile region, which are essential to address the demand, supply, and structural-induced scarcity challenges.

Water regime formation and capacity in the NRB

The Nile has been subject to numerous water-related disputes and confrontations over disputed governance (Sinnona, 2012, p. 3). According to Paisley and Henshaw (2013, p. 1), it faces a complex array of governance challenges. The continued mismanagement of the Nile and its disproportionate distribution increases the likelihood of tensions and disputes. Therefore, protecting, managing, and sustaining water and watersheds through proper water use is critical for every state, company, and community.

Garretson (1960, p. 144) argued that an NRB authority or administration is essential to addressing the problems relating to the full development and use of the NRB. According to Kukk and Deese (1996, p. 22), river basin organisations are essential for all riparian countries to voice and resolve water issues without resorting to force.

Without these organisations, political tensions and disputes along shared rivers tend to emerge (Kukk and Deese, 1996, p. 22). This is evident in the case of the NRB. The depletion, degradation, and political tensions and disputes in the NRB are due to the absence, or lack of capacity and capability, of a river basin organisation that mediates and ensures cooperation in the rational planning, conservation and development of the resources of the basin as a whole (Godana, 1985, p. 264).

There hasn't been much successful regime formation or regime success in the NRB. Regime formation in the Nile region has been limited due to a number of factors, which have also limited their successes. One of these factors is the lack of inclusive participation by Nile riparian countries, solely because of the dispute about the rights over, utilisation of, or both of the Nile. An example of this is UNDUGU (*brotherhood* in Swahili), which was formed to provide a forum for information sharing and to serve as an institution for sharing expertise and further allowing the riparian countries to become accustomed to treating the Nile as a whole, and not as less than the sum of its national parts (Paisley and Henshaw, 2013, p. 6).

Despite the meaning attached to UNDUGU, it didn't appeal to many riparian states as a form of brotherhood. UNDUGU was formed in 1983 at Egypt's request and included all Nile riparian countries except Kenya and Ethiopia, which participated as observers. The goal of UNDUGU was to foster economic, social, cultural, and technical understanding. Although individual Nile riparian countries may have been motivated to participate for varying reasons, some argued that while many of them were interested in fostering self-reliance and African interdependence, for Egypt, UNDUGU was a means to further extend its hegemonic influence across the Nile Basin (Paisley and Henshaw, 2013, p. 6).

Today, the Nile has become a recipe for conflict in the region over the demand, degradation, and distribution of water resources. The fact that the Nile, as an internationally shared watercourse, moves across state boundaries and is a source of life for surrounding communities, yet is unevenly distributed, raises numerous red flags for the potential of conflict (Urdal, 2005, p. 419). It is in this context that we need to critically analyse the role of a regional authority as a critical remedy.

The AU and the NBI represent possible mechanisms for the sustainability, management, and use of the Nile. On the one hand the AU, as a regional authority that has as its central role a focus on resolving issues detrimental to peace and stability on the continent, needs critical analysis as a successful mechanism of mediation.

Rahman (2011) argued that in the past, a number of ethnic and tribal issues, land disputes, and acquisitions in which the AU could easily mediate have dominated Africa's interstate conflicts.

On the other hand, the NBI poses as a successful mechanism on the basis that it is an inter-riparian commission aimed at coordinating both upstream and downstream interests. According to Lamere (2012), the NBI is the first contemporary attempt to forge a collective water-sharing agreement. It is the only legal framework for regional environmental management covering all 11 Nile riparian countries. Therefore, the NBI is of critical importance to this chapter.

The NBI's vision further makes it an ideal and important regime to address the use and management of the Nile. Its vision states,

> to achieve sustainable socioeconomic development through the equitable utilisation of, and benefits from, the common Nile Basin water resources.

Its characterisation of the NRB's water resources as *common* refutes the notion that Egypt and Sudan have priority use of the entirety of the basin's flow (Stoa, 2014, p. 1358).

According to Godana (1985, p. 264), basin states can only gain through the creation of a comprehensive commission serving as an institutional vehicle for cooperation. Nonetheless, Jacobs (2012, p. 120) considered it worth noting that multilateral cooperative arrangements are not the only solid footing by which cooperative norms can be achieved. They pointed out that the institutional design of multilateral agreements can either help or hinder how actors behave as a collective.

NBI

The changing geopolitical environment has influenced the development of interstate mechanisms to deal with potential conflict over water in volatile transnational water resources. A number of regional frameworks have been established in recent years to address environmental conflicts within transboundary basins. For instance, the Lake Victoria Basin Commission undertook to develop the Lake Victoria Environmental Management Project, which was designed to restore the degraded lake ecosystem, manage shared water and watersheds, and point out source pollution (Mbaiwa, 2004, p. 1323).

Several initiatives were proposed for the NRB; however, they were not inclusive, institutionalised, and structured mechanisms that could bring together all riparian countries to address a common agenda. According to Wolf and Newton (2008), from 1967 to 1992, the UNDP supported the Hydromet Project designed to collect hydrometeorological information within the basin.

Based on the findings of this project, the Technical Cooperation Committee for the Promotion of the Development and Environmental Protection of the Nile Basin (TECCONILE) was formed in 1993 with the idea of creating an informal dialogue between riparian nations. TECCONILE later resulted in the creation of the NRB Action Plan in 1995 to ensure joint planning in the use and development of the Nile's water (Wolf and Newton, 2008).

In 1997, the World Bank allowed the Council of Ministers of Water Affairs of the Nile Basin States (Nile-COM) to direct and coordinate donor activities within the basin, which led to Nile-COM working in cooperation with organisations such as the UNDP, the World Bank, and the Canadian International Development Agency (CIDA) (Wolf and Newton, 2008).

According to Paisley and Henshaw (2013, p. 9), the major stumbling block was, and continues to be, the issue of water security and existing treaties. For instance, Egypt and Sudan still argue based on the 1959 Nile Agreement. Despite the failure to resolve this issue, the NBI was established. However, water security has so far proven to be a significant impediment to moving beyond the transitional NBI to a permanent agreement. As a result of these stumbling blocks, in February 1999, the water ministers of nine of the then-ten Nile riparian countries agreed to replace TECCONILE with the NBI.

This signalled a major step by riparian countries to move to a more inclusive and comprehensive cooperation platform. The NBI[8] had the task of managing a complex river system faced with conflictual hydropolitics (Guvele, 2003; Othieno and Zondi, 2006, p. 1). The NBI is a ten-member regional intergovernmental partnership, with Eritrea holding observer status. These members include

Burundi, Egypt, Ethiopia, Kenya, Rwanda, Tanzania, Uganda, the DRC, Sudan, and the newly founded South Sudan (Balleh, 2014).

Nile-COM is the highest decision and policymaking body. It is composed of the ministers in charge of water affairs in each of the NBI member states. This decision-making body receives technical support and advice from the Nile Technical Advisory Committee on matters related to the cooperative management and development of the common NRB water resources. The advisory committee is comprised of senior governmental officials, two from each of the member states (NBI, 2013).

As a regional intergovernmental partnership, the NBI seeks to develop the Nile cooperatively and establish a comprehensive framework to plan, manage, and use the natural resources of the Nile, principally water (Oloo, 2007, p. 95). The NBI also aims to include all members through participatory processes of dialogues (Wiebe 2001, p. 751; World Bank, 2008, p. 2). The initiative was launched with the understanding that a cooperative effort in developing and managing the Nile would bring the greatest level of mutual benefit to the region. It was through this inclusivity that an understanding of the implications of member states' actions on neighbours, and opportunities for managing risks and realising tangible benefits were created (Balleh, 2014).

The NBI is determined to oversee the cooperative, substantial socioeconomic benefits and regional peace and security framework of the NRB (Nile Basin Initiative, 2014). However, the NBI is faced with challenges detrimental to its effectiveness within the region. These include the challenges of transforming the NBI into a permanent Nile River Basin Commission (NRBC); the stalemate in its Cooperative Framework Agreement (CFA), which aims to replace the colonial-era treaty that gave Egypt and Sudan a majority share of the *Nile's water*; third party involvement and foreign aid dependence; and unilateralism in policy decision-making. These challenges are discussed in the following sections in detail.

Challenges to transforming the NBI into a Commission

The NBI is currently serving in the capacity of an interim organisation that has the authority and means to facilitate a more permanent legal and institutional arrangement to regulate the Nile (NRBC) (Kassa, 2013). As such, the NBI is just a transitional mechanism that is expected to phase out when a permanent NRBC is established following the conclusion of the CFA (Balleh, 2014). The CFA establishes the principles, operational mechanisms, and setup of the NRBC to facilitate and oversee the smooth equitable and reasonable use, management, and protection of the Nile's water. The CFA should be included in national legislation; therefore, disbanding the NBI and simultaneously establishing a permanent NRBC (Lie, 2010, p. 11).

For the initiative to convert into a commission, 6 of the 11 countries' parliaments must ratify the CFA as law. Ethiopia and Rwanda have done so. Tanzania, Burundi, and South Sudan are waiting for ratification. The Ugandan parliament is expected to vote on it soon. In Kenya, the law is at the cabinet level (Tekle, 2017). In short, the NBI needs a legal status to become a permanent NRBC (Lie, 2010, p. 2).

Transforming the initiative into a commission entails a different level of competencies. For instance, when the initiative converts into a commission, it will secure financial support from its development partners and will enable financial support by having its revenue generation scheme. The NRBC will ensure the use of the Nile by member states on a legal basis (Kassa, 2013).

The establishment of an interstate commission (the NRBC) complements article V(3) of the AU convention on the conservation of nature and natural resources. This constitutes a departure from the past trend of unilateral water use and management, particularly when emphasising the issues of fair water allocation, joint management, and developing resources (Lie, 2010, p. 11). This also signals a fundamental shift in the status quo of upstream–downstream relations, particularly in the relationship between Egypt, Sudan, and Ethiopia (Lie, 2010, p. 11).

So far, the NBI has not achieved its goal of being transformed into a commission. The CFA faces a stalemate due to conflicting arguments between upstream and downstream riparian countries regarding some of its provisions; therefore, delaying the establishment of a basin commission (Waslekar, 2013, p. 7) (see the section entitled *Dispute over the cooperative framework agreement* in the following section).

Since a basin commission has not been established, the basin is currently administered by the NBI. However, failing to establish a commission and having to deal with an initiative raises challenges for the full development and use of the Nile. For instance, an initiative has no legal status and, as a result, it cannot make binding and enforceable decisions. It can only enjoy financial support from development partners. This means having to rely on unconventional donors. Moreover, the impasse experienced in the CFA that is supposed to transform the NBI into an NRBC has threatened this support from development partners, which makes future support uncertain (Balleh, 2014).

Dispute over the cooperative framework agreement

The CFA is an important example of a multilateral agreement that hinders how actors behave as a collective. The NBI proposed a CFA that would include all the Nile riparian countries and incorporate the principles, structures, and institutions of the NBI. The CFA aims to provide legal principles to determine reasonable and equitable solutions for sharing the Nile among the basin states and to guarantee the water security of the Nile riparian countries (Lie, 2010, p. 11; Salman, 2013, pp. 20–21).

The articles of the agreement include the general aims and principles of protection, equitable and reasonable use, prevention of significant harm, community interest, data and information exchange, peaceful resolution of disputes, conservation, and sustainable development of the Nile (Jacobs, 2012, p. 121; Salman, 2013, p. 21).

Since its formation, the agreement has faced major complexities. The status of the CFA is disputed among the Nile riparian states. Lino (2013, p. 13) asserted,

> These complexities manifest themselves in the failure, thus far, of the basin countries to reach an agreement on an inclusive CFA, more than ten years after the start of the facilitative efforts under the NBI.

These difficulties are a result of the respective positions of the upstream countries on the colonial treaties, as well as the Egyptian and Sudanese claims to what they see as their acquired uses and rights of the Nile. Even the ministerial meetings that took place in Kinshasa, Alexandria, and Sharm El-Sheikh in 2009 and 2010 failed to resolve these differences (Salman, 2013, p. 20), leaving the CFA ineffective and in a stalemate.

The CFA owes much of this stalemate first to the contradiction and controversy caused by its principles, and second to the conflicting positions regarding rights and provisions on the use of the Nile, particularly in Articles 4 and 14(a) (Waslekar, 2013, p. 7).

Salman (2013, p. 21) argued that the agreement causes contradictions and controversy between upstream and downstream countries by including the principles of *prevention of causing significant harm to other basin states* and *equitable and reasonable use*. The latter entails:

> The state whose use causes harm to other basin states, in the absence of agreement to such use, take all appropriate measures, having due regard for the provisions of the CFA on equitable and reasonable utilisation, in consultation with the affected state, to eliminate or mitigate such harm and, where appropriate, to discuss the question of compensation.

As a general rule, upstream countries are likely to favour the equitable use principle because it provides more scope for states to use their share of the watercourse for activities that may impact downstream countries, while downstream countries will favour the no-harm principle since it protects their existing uses against impacts resulting from activities undertaken by upstream countries (Salman, 2013, p. 21).

According to Article 4:

> [The] Nile Basin States shall in their respective territories utilize the water resources of the Nile River system and the NRB in an equitable and reasonable manner. In particular, those water resources shall be used and developed by Nile Basin States with a view to attaining optimal and sustainable utilization thereof and benefits there from, taking into account the interests of the Basin States concerned, consistent with adequate protection of those water

> resources. Each Basin State is entitled to an equitable and reasonable share in the beneficial uses of the water resources of the Nile River system and the NRB,
>
> (International Water Law Project, 2014a)

Article 4 is essentially about *equitable and reasonable use* of the Nile. Article 4 considers the equitable and reasonable use of both the water resources of the *Nile River System* and the *NRB*. What this means, respectively, is that the Nile River System allows for the equitable and reasonable use of *only the water that runs into the river*, while the NRB refers to and allows for the equitable and reasonable use of *all the water resources of the Nile* (Al-Ahram weekly, 2015).

> Nile basin states agree in the spirit of cooperation, not to significantly affect the water security of any other Nile basin state
>
> (Mbaku and Mwangi, 2015)

Egypt and Sudan on the other hand, want Article 14 amended to reflect their historic rights. Based on Article 14, Egypt and Sudan argue,

> Nile basin states agree, in a spirit of cooperation, not to adversely affect the water security and current uses and rights of any other Nile basin states
>
> (*Sudan Tribune*, 2008; Salman, 2013, p. 21)

Egypt required the CFA to acknowledge the 1959 Nile Agreement rather than supersede it (Lie, 2010, p. 11). This has resulted in the current stalemate in the NBI cooperative framework. The danger of not agreeing on the CFA is the threat to the emergence of the NRBC as indicated earlier. In this way, the institutional design of the CFA has hindered how actors behave as a collective as far as its articles are concerned.

In addition, the upstream–downstream position causes conflict for the CFA with claims of historical and natural rights over the use of the Nile (Kendie, 1999, p. 145). On the one hand, upstream countries have insisted this framework must disregard all previous agreements of which they were not part. Downstream countries seek a new framework to incorporate such earlier agreements. This has caused more disparities within the region (Oloo, 2007, p. 101). The challenge was acknowledged by Beyene when he stated,

> an important challenge the Nile Basin countries are facing is related to establishing a permanent legal and institutional set-up to anchor their cooperation on a solid footing,
>
> (Rutagwera, 2014)

This means that the agreement, once effective, will transform the NBI into a permanent NRBC (Steinbrueck, 2014, p. 11). According to Othieno and Zondi (2006, pp. 2–3):

> The agreement is also expected to provide a resolution to such issues pertaining to the balance between who has the historic right to the Nile, and issues of sovereign water rights especially with regards to where the river ought to be controlled, whether that be upstream or downstream.

This will eventually allow the status quo governing Nile water treaties to be rectified and enable the Nile riparian countries to develop a common position on the equitable exploitation and use of NRB resources. In this instance, the agreement has not addressed the equitable and reasonable use of the Nile, which would resolve the disputes between upstream and downstream riparian countries and develop a community of interest.

Third-party involvement and foreign aid dependence

Although the AU has expressed interest in endorsing the NBI, it is unclear whether this stance emboldens or deters the World Bank, the International Monetary Fund, or other foreign donors from getting involved (Ferar, 2010, pp. 91–92). The NBI faces similar challenges in terms of involvement and funding by foreign donors (Foulds, 2002, p. 2). It is evident that foreign donors (e.g., the World Bank and CIDA) have been involved in the funding process that led to the establishment of the initiative (Wolf and Newton, 2008).

These are the same donors that obscured the cooperation framework in the NRB and, therefore, dictate the agenda of water cooperation in favour of technical rather than political interventions such as preventative diplomacy. These technical interventions, for instance, include the World Bank's intervention to stop Ethiopia from building the GERD, or the refusal of funding for building this project. This ultimately protects Egyptian hydro-hegemony (Othieno and Zondi, 2006, p. 3).

Executive Director Engineer Teferra Beyene has also acknowledged the financial challenges faced by the initiative. However, instead of Beyene motivating riparian member states to take financial responsibility for the initiative rather than depending on foreign aid, he acknowledges the latter. This is harmful to policymaking since third parties are constantly involved and influence important decisions of the initiative.

During the Chinese delegation on a China–Africa transboundary water study tour on 7 May 2014, Beyene stated:

> The biggest challenge the NBI faces now, is lack of finance to implement the investment projects it has prepared and ensuring its institutional sustainability to continue providing the excellent services it has been providing for its member states. I believe the partnership that NBI has recently started with the Republic of China will help in filling some of these gaps as China is a key partner in Africa's bid.
>
> (Rutagwera, 2014)

This statement signifies a deeper entrenchment into foreign aid dependence.

Unilateralism in policy decision-making

The NBI faces challenges in addressing the unilateralism that exists in decision-making processes and laws governing the Nile. As has been argued, Egypt as the hydro-hegemony dominates and dictates these decision-making processes, which therefore remain in favour of Egyptian interests. This is evident in the CFA where the initiative has failed to resolve the stalemate due to conflicting rights at the expense of win-win gains. Progress on a cooperative framework constantly fails to materialise due to Egypt's disapproval. Therefore, these excessive unilateral actions pose a challenge to achieving mutually beneficial cooperation (Link *et al.*, 2014, pp. 4–9).

It has been argued that the failure of the NBI to address unilateralism in its policy decisions lies in its having involved only water technocrats in the process, rather than invoking an institutionalised political will from the riparian states. This reduces it to a mere water authority. It lacks a political centre and a presence in the public eye of participating states. Without this political will, the issue of the equitable use of the Nile is impossible and faces a number of political compromises (Oloo, 2007, pp. 101–104).

This unilateralism prompts competition, leading to lose-lose gains and the failure of the NBI. The inability to establish the NRBC will exacerbate unilateral decision-making by riparian states regarding major water resource investment projects (hydropower projects) and, therefore, increase the level of disputes in the region.

Conclusion

The shortage of water from other water users has the potential to cause violent conflict (Homer-Dixon, 1999a, p. 48). Since China's growth in upstream countries' hydropower development is inevitable, this book argues that there needs to be an adequate balance between water demand management and a supply-oriented approach with the help of regimes in the region to avoid the emergence of disputes within the region (GeeskaAfrica Online, 2015). It can therefore be argued that given the lack of consensus and legitimacy of these agreements, a coordinated approach to the use and management of the Nile's water under the aegis of continental and basin-wide regimes is imperative.

The NBI has played a limited role in the water diplomacy efforts between the upstream and downstream countries. Both regimes have respectively failed to address demand and supply-induced scarcities and to clarify the validity, invalidity, or both of the Nile water agreements, which are the cause of increased potential for conflict in the basin. This inability by the AU and NBI is detrimental to peace and security in the NRB, because it has resulted in severe pressure from uncoordinated water use and management.

It is important to note that any institution, organisation, or commission entrusted with the responsibility of ensuring the management and distribution of the Nile's water is, in one way or the other, required to address the salient features of environmental scarcity that challenge stability in the NRB. It, therefore, needs to consider the interdependence of these forms of scarcity. With the continued existence of these challenges in critical bodies entrusted with the responsibility of developing a cooperative solution in the region, the potential for conflict in the NRB remains (Othieno and Zondi, 2006, p. 3).

To avoid conflict in the basin, the AU and BI must clarify that the Nile is a shared resource and that what happens in one country, negative or positive, affects the others: hence the need for joint management and development for the benefit of all. Nonetheless, despite the AU and NBI having shared watercourse agreements, they do not seem to know how to deal with environmental scarcity issues. According to Arnold (2013), this has led to many believing that one possible way to resolve the Nile dispute could be to involve the ICJ, similar to how the Okavango disputes were resolved.

Notes

1 Egypt
2 Both in Sudan, north of the capital Khartoum
3 Both in Ethiopia
4 Ethiopia
5 Ethiopia
6 Kenya
7 Located in east central Africa along the equator and borders the countries of Uganda, Kenya, and Tanzania
8 Also referred to as the Entebbe Treaty or Entebbe Agreement because it was signed and is based in Entebbe, Uganda; the NBI chair rotates among its members (Lie, 2010, p. 13)

Bibliography

Abseno, M.M. (2013) 'Nile river basin', in Loures, F.R. and Rieu-Clarke A. (eds.) *The UN watercourses convention in force: Strengthening international law for transboundary water management*. Abingdon: Routledge.

Adar, K.G. (2007) 'Kenya's foreign policy and geopolitical interest: The case of the Nile River Basin', *African Sociological Review*, 11(1).

Adar, K.G., 2011. 'Kenya's foreign-policy and geopolitical interests: The case of the Nile River Basin', in Adar, K.G. and Check, N.A. (eds.) *Cooperative diplomacy, regional stability and national interests: The Nile River and the riparian states*. Pretoria: Africa Institute of South Africa.

Afedraru, L. (2014) 'Uganda's energy sector grows despite challenges', *Daily Monitor*, 26 November. Available at: http://mobile.monitor.co.ug/Business/technology/Uganda-s-energy-sector--grows-despite-challenges/-/2471068/2535688/-/format/xhtml/-/13yla7j/-/index.html. (Accessed: 4 April 2015).

African Union. (2008) Statement by H.E. Professor Alpha Oumar Konare Chairperson of the commission of the African Union, The occasion of the six anniversary of Africa environment day, Addis Ababa, 3 March 2008.

Al Jazeera. (2022) Ethiopia's Blue Nile mega-dam explained. 21 February. Available at: www.aljazeera.com/news/2021/7/8/explainer-ethiopias-massive-nile-dam. (Accessed: 27 July 2022).

Al-Ahram Weekly. (2015) 'Questions about the CFA', 22 April. Available at: http://weekly.ahram.org.eg/News/11030/17/Questions-about-the-CFA-.aspx. (Accessed: 19 January 2017).
Al-Din Amer, S. (1997) 'The law of water: Historical record', in Dupuy, B. (ed.) *Aspects économiques de la gestion de l'eaudans le bassinméditerranéen*. Paris. International Centre for Advanced Mediterranean Agronomic Studies
Allan, A.J. (1999) 'The Nile Basin: Evolving approaches to Nile waters management', *SOAS Water Issues Group*: England, University of London.
Anna, C. (2020) 'Satellite images show Ethiopia dam reservoir swelling', *AP News*. 14 July 14. Available at: https://apnews.com/article/ethiopia-sudan-africa-international-news-egypt-600710323bb9bb4427a3d3b9a3115bae. (Accessed: 6 August 2022).
Arnold, D. (2013) 'Nile power shifts away from Egypt', *VOA.news*. 15 July. Available at: www.voanews.com/content/nile-power-shifts-away-from-egypt/1702289.html. (Accessed: 17 February 2015).
Ashton, P. J. (2002) 'Avoiding conflict over Africa's resources', AMBIO, *A Journal of the Human Environment*, 31(3).
Ayad, A.R. (2013) *Water quality in Cairo, is it safe?', Cairo from Below: Information and debate on the urban future of Cairo*, 5 February. Available at: http://cairofrombelow.org/2013/02/05/water-quality-and-cairo-is-it-safe/. (Accessed: 18 January 2014).
Baecher, G. *et al.* (2000) 'The Nile Basin: Environmental transboundary opportunities and constraints analysis', United States Agency for International Development Draft. Harvard Institute for International Development (HIID): Harvard University. Cambridge, Massachusetts.
Balleh, A. (2014) 'Nile Basin Initiative: Has it delivered on its promises?' *Horn Affairs*, 11 September, Available at: http://hornaffairs.com/en/2014/09/11/nile-basin-initiative-nbi-has-it-delivered-on-its-promises/. (Accessed: 12 November 2014).
Basheer, M. *et al.* (2020) 'Filling Africa's largest hydropower dam should consider engineering realities', *One Earth*, 3(3).
BBC News. (2010) 'Web campaign against Ethiopia Gibe III dam', 23 March, Available at: http://news.bbc.co.uk/2/hi/africa/8582682.stm. (Accessed: 20 March 2015).
BBC News. (2022) 'Ethiopia starts generating power from River Nile dam', 20 February, Available www.bbc.com/news/world-africa-60451702. (Accessed: 6 August 2022).
Bigagaza, J., Abong, C., and Mukarubuga, C. (2002) 'Land scarcity, distribution and conflict in Rwanda', in Lind, J. and Sturman, K. (eds.) *Scarcity and surfeit: The ecology of Africa's conflicts*. Nairobi and Pretoria: African Centre for Technology Studies Press and Institute for Security Studies.
Bingham, K. (2001) 'Environmental scarcity and the outbreak of conflict', Population Reference Bureau Inform (PRB), January. Available at: www.prb.org/Publications/Articles/2001/ EnvironmentalScarcityandtheOutbreakofConflict.aspx. (Accessed: 29 December 2013).
Bosshard, P. (2009) 'China dams the world', *World Policy Journal*, 26(4).
Bosshard, P. (2015). 'Chinese Dams in Africa', *International Rivers*. Available at: www.internationalrivers.org/campaigns/chinese-dams-in-africa. (Accessed: 6 March 2015).
Braune, E and Xu, Y. (2010) 'The role of ground water in sub-Saharan Africa', *Ground Water Issues*, 48(2).
Brewer, N., (2008). 'The new great walls: A guide to China's overseas dam industry', International Rivers, 15 July. Available at: www.internationalrivers.org/resources/the-new-great-walls-a-guide-to-china%E2%80%99s-overseas-dam-industry-3962. (Accessed: 16 May 2017).

Caponera, D. (1993). 'Legal aspects of transboundary river basins in the Middle East: The Al Asi (Orontes)', *Jordan and the Nile, Journal of Natural Resources*, 33.

Cascão, A.E. (2009) 'Changing power relations in the Nile River basin: Unilateralism vs. cooperation?' *Water Alternatives*, 2(2).

Chesire, K.D. (2010) *Control over the Nile: Implications across Nations*. Monterey, California: Naval Postgraduate School.

Conca, K., Ehrlich, A.H. and Gleick, P. (2000) 'Conflict over resources', Draft Background Paper for Working Group 5: Resources and Environmental Degradation as Sources of Conflict', *50th Pugwash Conference on Science and World Affairs: Eliminating the Cause of War*. Queens' College, Cambridge, 3–8 August.

Conniff, R. (2017) 'The vanishing Nile: A great river faces a multitude of threats', *Yale Environment*. 360. Available at: http://e360.yale.edu/features/vanishing-nile-a-great-river-faces-a-multitude-of-threats-egypt-dam. (Accessed: 16 May 2017).

Dakkak, A. (2013) 'Egypt's water crisis – Recipe for disaster', *EcoMENA: Echoing Sustainability*, 30 March. Available at: www.ecomena.org/egypt-water/. (Accessed: 18 January 2014).

Degefu, G. (2003). *The Nile: Historical, legal and developmental perspectives*. Victoria, Canada: Trafford Publishing.

Demeke, M.A. (2013) 'Why Ethiopia rejected the validity of the 1929 and 1959 agreements regarding the Nile River', *The Reporter*, 22 June. Available at: www.thereporterethiopia.com/index.php/opinion/commentary/item/627-why-130ethiopia-rejected-the-validity-of-the-1929-and-1959-agreements-regarding-the-nile-river. (Accessed: 12 September 2014).

Deng, B.K. (2007) 'Cooperation between Egypt and Sudan over the Nile River waters: The challenges of duality', *African Sociological Review* , 11(1).

Desta, H. (2012) 'Overview on degradation of land due to overgrazing in Ethiopia', Department of Veterinary Medicine and Animal Sciences, University of Milan.

Dunoff, J.L., and Trachtman, J.P(1999) 'Economic analysis of international law', *Yale Journal of International Law*, 1.

El-Fadel, M., El-Sayegh, Y., El-Fadl, K., and Khorbotly, D. (2003) 'The Nile River Basin: A case study in surface water conflict resolution', *Journal for Natural Resource Life Science Education*, 32.

El-Sayed, H. *et al*. (2020) 'The Nile water-food-energy nexus under uncertainty: Impacts of the grand Ethiopian renaissance dam. *Journal of Water Resources Planning Management*, 146(11).

Ferar, K. (2010). Nile water: Development, environment and conflict. Unpublished dissertation. Claremont: Pitzer College.

Few, R. *et al.* (2015) 'Vulnerability and adaptation to climate change in semi-arid areas in East Africa', Adaptation at Scale in Semi-Arid Regions (ASSAR). South Africa: ASSAR Project Management Unit (PMU).

Di Nunzio, J. (2013) 'Conflict on the Nile: the future of transboundary water disputes over the world's longest river', Strategic Analysis Paper, 25 November, Future Directions International: Nedlands, Australia.

Foulds, K. (2002) 'The Nile Basin Initiative: Challenges to implementation', *Managing Shared Water Conference*. Hamilton, Ontario. Canada. 23–28 June 2002.

Garretson, A.H. (1960) 'The Nile River system', *Proceedings of the American Society of International Law*, 54.

GeeskaAfrika. (2015) 'Ethiopia: Global Dam's water transfer not enough for security.' Available at: www.geeskaafrika.com/ethiopia-global-dams-water-transfers-not-enough-for-security/7659/. (Accessed: 20 March 2015).

Godana, B.A. (1985). *Africa's shared water resources*. London: Frances Pinter: Boulder, Colorado: Lynne Rienner Publishers.

Green Prophet. (2010) 'Upstream African countries sign Nile Water deal against Egypt's interests.18 May. Available at: www.greenprophet.com/2010/05/upstream-african-countries-sign-nile-water-deal-against-egypts-interest/. (Accessed: 30 October 2013).

Greste, P. (2009) 'The dams that divides Ethiopians', BBC News, 26 March. Available at: http://news.bbc.co.uk/2/hi/africa/7959444.stm. (Accessed: 12 March 2015).

Guvele, A.C. (2003) 'The Nile Basin Initiative and its implications in Post Conflict South Sudan', Unpublished research paper Available at: http://cafnr.missouri.edu/iap/sudan/doc/nile-basin.pdf. (Accessed: 22 April 2014).

Haftendorn, H. (1999) 'Water and International Conflicts', *International Studies Association Annual Convention*, Washington, D.C., 16–20 February.

Hassan, H.A. and Rasheedy, A.A. (2011) 'The Nile River and Egyptian foreign-policy interests', in Adar, K.G. and Check, N.A. (eds.) *Cooperative Diplomacy, Regional Stability and National Interests. The Nile River and the Riparian States*, Pretoria: Africa Institute of South Africa, pp. 131-52.

Hegazi, F.F. (2011) 'Cooperation over water in the Eastern Nile Basin: Obstacles & opportunities. Unpublished dissertation. Duke University.

Homer-Dixon, T.F and Blitt, J. (1998) *Ecoviolence: Links Among Environment, Population, and Security*. New York: Rowman and Littlefield Publishers.

Homer-Dixon, T.F. (1994) 'Environmental scarcities and violent conflict: Evidence from cases', *International Security*, 19,(1).

Homer-Dixon, T.F. (1999a). *Environment, Scarcity, and Violence*. Princeton University Press: New Jersey.

International Water Law Project (2014a) 'Agreement on the Nile River Basin Cooperative Framework. Available at: http://internationalwaterlaw.org/documents/regionaldocs/Nile_River_Basin_Cooperative_Framework_2010.pdf. (Accessed: 14 September 2014).

International Water Law Project. (2014b) 'Addressing the future of water law and policy in the 21st century'. Available at: www.internationalwaterlaw.org/documents/%20regionaldocs/Nile_River_Basin_Cooperative_Framework_2010.pdf. (Accessed: 14 September 2014).

Jacobs, I.M. (2006) *The water war debate: Swimming upstream or downstream in the Okavango and the Nile*. Unpublished dissertation. University of Stellenbosch.

Jacobs, I.M. (2012) 'The politics of water in Africa: Norms, environmental regions and transboundary', *Cooperation in the Orange-Senqu and the Nile Rivers*. New York: Bloomsbury Academic.

Kameri-Mbote, P. (2007) 'Water, Conflict, and Cooperation: Lessons from the Nile River Basin', Navigating Peace, No.4, *Environmental Change and Security Program*, Woodrow Wilson International Center for scholars, January.

Karyabwite, D.R. (2000) Water Scarcity in the Nile River Valley, United Nations Environment Programme: Geneva, Switzerland. Available at: www.grid.unep.ch/activities/sustainable/nile/nilereport.pdf (Accessed: 12 February 2017).

Kassa, T. (2010) 'International law and moderations of physical geography: The Nile setting', in Tvedt, T. Chapman, G. and Hagen, R. (eds.) *Water, geopolitics and the new world order. A history of water*. Series II Volume 3.I. London: B. Tauris,

Kassa, Z. (2013) 'Ethiopia endorses Nile Basin Cooperative Framework Agreement', *AllAfrica*, 13 June, Available at: www.allafrica.com/stories/201306141038.html. (Accessed: 12 February 2014).

Katz, C.L. (2013) 'Another cup at the Nile's crowded Spigot: South Sudan and its Nile water rights', *Georgetown Journal of International Law*, 44

Kendie, D. (1999) 'Egypt and the hydro-politics of the Blue Nile River', *Northeast African Studies*, 6(1-2).

Kieyah, J. (2007) *The Nile Waters Agreement: Legal and economic Analysis. from selected works of Joseph Kieyah*. Nairobi: Kenya Institute for Public Policy Research and Analysis.

Kigomo, B.N. (2003) 'Forests and woodlands degradation in dryland Africa: A case for urgent global attention'. Available at: www.fao.org/docrep/article/wfc/xii/0169-b3.html. (Accessed: 14 September 2014).

Kim, U., and Kalaurachchi, J.J. (2009) 'Climate change impacts on water resources in the upper Blue Nile River basin, Ethiopia', *Journal of the American Water Resources Association*, 45(6).

Klare, M.T. (2002) *Resource wars: The new landscape of global conflict*. New York: Henry Holt and Company.

Kliot, N., Shmueli, D. and Shamir, U. (2001) 'Institutions for management of transboundary water resources: their nature, character and shortcomings', *Water Policy*. 3.

Knobelsdorf, V. (2006) 'The Nile Waters Agreements: Imposition and impacts of a transboundary legal system', *Columbia Journal of Transnational Law*, 44(2).

Kukk, C. and Deese, D. (1996) 'At the water's edge: Regional conflict and cooperation over fresh water', *UCLA Journal of International Law and Foreign Affairs*, 21(1).

Kwetegyeka, J. *et al*. (2014) 'Impact of the disused Kilembe mine pyrites on the domestic water quality of Kasese town, Western Uganda', *Caribbean Journal of Science and Technology*, 2.

Lamere, C. (2012) 'Nile Basin at a turning point as political change roil balance of power and competing demands proliferate', *New Security Beat*, 4 September. Available at: www.newsecuritybeat.org/2012/09/nile-basin-turning-point-egyptian-revolution-roils-balance-power-competing-demands-proliferate/. (Accessed:20 May 2014).

Libiszewski, S. (1992) 'What is an Environmental Conflict?' *First coordination meeting of Environment and Conflict Project (ENCOP)*, Zurich. 30 April–1 May.

Lie, H.S. (2010) 'Supporting the Nile River Basin: A political analysis beyond the river', Norwegian Institute of International Affairs.

Link, M.P., Piontek, F., Scheffran, J. and Schilling, J. (2014) *On foes and flows: Water conflict and cooperation in the Nile River Basin in times of climate change*. Grindelberg: University of Hamburg.

Lino, V. (2013) *Post-independence South Sudan: Prospects and implication on the Nile basin hydro-politics*. Unpublished dissertation. University of Nairobi.

Lowi, M.R. (1995) 'Rivers of conflict, rivers of peace', *Journal of International Affairs*, 49(1).

Lumumba, P.L.O. (2007) 'The interpretation of the 1929 Treaty and its legal relevance and implications for the stability of the region', *African Sociological Review*, 11(1).

Luscombe, S. (2012) *19th Century Timeline*. Available at: www.britishempire.co.uk/timeline/19century.htm. (Accessed: 20 May 2014).

Marterns, A.K. (2011) 'Impacts of Global Change on the Nile Basin: Options for Hydropolitical Reform in Egypt and Ethiopia', *International Food Policy Research Institute (IFPRI)*.

Marty, F. (2001) *Managing international rivers: Problems, politics and institutions*. Bern: Peter Lang.

Mayton, J. (2012) 'Water for all: The Nile Basin conundrum', *DANTEmag*, 8 December. Available at: www.dantemag.com/2012/12/water-for-all-the-nile-basin-conundrum/. (Accessed: 30 October 2013).

Mbaku, J. M and Mwangi, S. K. (2015). 'The limits of the new "Nile" agreement'. *Africa in Focus*, 28th April. Available at. www.brookings.edu/blog/africa-in-focus/2015/04/28/the-limits-of-the-new-nile-agreement/. (Accessed: 16 May 2017).
McCaffrey, S.C. (2003) *The Law of international watercourses*. 2nd edition. Oxford: Oxford University Press.
Moffet, G.D. (1990) 'Nile River rights run deep', *The Christian Science Monitor*, 13 March. Available at: www.csmonitor.com/1990/0313/o1wat2.html. (Accessed: 23 February 2015).
Muramuzi, F. (2011) 'Sustainable development', Aljazeera, 13 June. Available at: www.aljazeera.com/programmes/struggleoverthenile/2011/06/2011612142731608591.html. (Accessed: 04 April 2015).
NBI. (2005) 'Nile basin regional water quality monitoring baseline study report for Burundi, DRC, Egypt, Ethiopia, Kenya, Rwanda, Sudan, Tanzania, and Uganda', *Nile Trans boundary Environmental Action Project.*
NBI. (2012a) *One River One People One Vision*. Entebbe: NBI.
NBI. (2012b) 'Nile information system: The state of the River Nile Basin.' Available at: http://nileis.nilebasin.org/system/file/Nile%20SoB%20Report%20Chapter %209%20-%20SUmmary.pdf. (Accessed: 12 November 2013).
NBI. (2013) 'Nile Basin Initiative Corporate Report, July 2012-June 2013'. Entebbe: NBI.
NBI. (2014) Nile Basin Initiative. Available at: http://nilebasin.org/index.php/ about-us/nile-basin-initiative. (Accessed: 2 June 2014).
Oestigaard, T. (2012) 'Water scarcity and food scarcity along the Nile: Politics, population increase and climate change', *Current African Issues*, 49.
Ohlsson, L. (1999) *Environment, scarcity and Conflict: A study of Malthusian concerns*, Göteburg: Department of Peace and Development Research.
Okascha, R. (2012) *Water scarcity and regional security in the Nile Basin*. Unpublished thesis. Phillips University of Marburg, Center for Near and Middle East Studies.
Okbazaghi, Y. (2008) *Water Resources and inter-riparian relations in the Nile basin: The search for an integrative discourse*. Albany: State University of New York Press.
Okoth-Owiro, A. (2004) 'The Nile Treaty: State succession and international treaty Commitments: A case study of the Nile water Treaties', Occasional Papers series. Nairobi: Konrad Adenauer Foundation.
Oloo, A. (2007) 'The quest for cooperation in the Nile Water Conflicts: The case of Eritrea', *African Sociological Review*, 11(1).
Othieno, T. and Zondi, S. (2006) 'Co-operative management of Africa's water resources: A basis for resolving conflicts on the Nile River Basin', *Institute for Global Dialogue*, 59(5).
Paisley, R.K. and Henshaw, T.W. (2013). *Transboundary governance of the Nile River Basin: Past, present and future. International waters governance initiative.* Vancouver: University of British Columbia.
Pakes, L. (2013) 'The politics of water scarcity in the Nile Basin: The case of Egypt', *Journal of Politics and International Studies*, 9.
Pottinger, L. (1996) 'Environmental impacts of large dams: African examples', *International Rivers*, 1(10). Available at: www.internationalrivers.org/en/ africa/environmentalimpacts-large-dams-african-examples. (Accessed: 4 April 2015).
Rahman, M.A. (2011) 'The geopolitics of water in the Nile River Basin', *Global Research*, 24 July. Available at: www.globalresearch.ca/the-geopolitics-of-water-in-the-Nile-river-basin/25746. (Accessed: 6 April 2013).

Reyskens, M. (2011) 'In hot pursuit: Resource wars in Africa', 16 November. Available at: www.consultancyafrica.com/index.php?option=com_content&view=article&id=901:in-hot-pursuit-resource-wars-in-africa-&catid=60:conflict-terrorism-discussion-papers&Itemid=265. (Accessed: 19 April 2013).

Rubenson, S. (1991) *The survival of Ethiopian independence*. Addis Ababa: Kuraz Publishing Agency.

Rutagwera, P. (2014) 'NBI and Chinese delegation discuss future partnerships', *International Waters Learning Exchange and Resource Network*, 8 May, Available at: http://iwlearn.net/iw-projects/3398/news/nile-basin-initiative-news/100d98783a7a69fdd757308fd765ab82. (Accessed: 25 June 2014).

Salman, S. (2013) 'The Nile Basin Cooperative Framework Agreement: A peaceful unfolding African Spring?' *Water International*, 38 (1).

Savenije, H.G. and Van der Zaag P. (2000) 'Conceptual framework for the management of shared river basins with special reference to the SADC and EU, *Water Policy* , 2(1–2).

Schilling, D.R. (2013) 'Egypt's $90 Billion South Valley Project', Industry Tap, 10 January. Available at: www.industrytap.com/egypts-90-billion-south-valley-project/539. (Accessed: 19 February 2015).

Scudder, T. (2003) 'The Aswan High Dam case', Unpublished Manuscript. Available at: http://people.hss.caltech.edu/~tzs/Aswan%20High%20Dam%20case.pdf. (Accessed: 4 April 2015).

Shinn, D. H. (2006) 'Nile Basin relations: Egypt, Sudan and Ethiopia, Elliott School of International Affairs, The George Washington University.Available at: http://elliott.gwu.edu/news/speeches/shinn0706_nilebasin.cfm. (Accessed: 6 February 2013).

Sinclair, M. (1996) 'The environmental cooperation agreement between Mexico and the United States: A response to the pollution problems of the Borderlands', *Cornell International Law Journal*, 19.

Sinnona, G. (2012). *Two-tier water governance in the Nile River Basin*. Available at: www.feem-web.it/ess/ess12/files/papers/sinnona.pdf. (Accessed: 4 June 2014).

Solomon, H. (1996). 'Towards the 21st century: A new global security agenda?' *Institute for Defence Policy*, 6.

Sombroek, W. and Sene, E. (1999) 'Land degradation in arid, semi-arid and dry sub-humid areas: rainfed and irrigated lands, rangelands and woodlands', *Inter-governmental negotiating committee for the preparation of a convention to combat desertification and drought (INCD), First Substantive Session*, Nairobi, 24 May–4 June.

Steinbrueck, E. (2014) *Water scarcity and international conflict in Africa: A game theory approach*. Water Resource and Protection: Swedish Water House. Stockholm, Sweden.

Stoa, R.B. (2014)'The United Nations Watercourses Convention on the dawn of entry into force', *Vanderbilt Journal of Transnational Law*, 47.

Sudan Tribune. (2008) 'Kenya says Sudan and Egypt want to monopolize Nile water', 9 November. Available at: www.sudantribune.com/spip.php?article29199. (Accessed: 14 September 2014).

Swain, A. (2011) 'Challenges for water sharing in the Nile basin: Changing geo-politics and changing climate', *Hydrological Sciences Journal*, 56(4).

Swain, A., and Jamali, Q. (2011) 'New challenges for Nile Basin cooperation', *New Routes* 15(3).

Tekle, T. (2017) 'ENTRO says Nile Basin Commission to be established', Sudan Tribune, 12 August. Available at: https://sudantribune.com/article61411/. (Accessed: 12 September 2017).

Thakur, M. (2009) 'Building on Progress? Chinese engagement in Ethiopia', *China in Africa Project occasional paper No. 38.South African Institute of International Affairs (SAIIA)*: Johannesburg: The University of the Witwatersrand.

The Journal. (2012) 'UN: Somali famine is over, but action still needed,' Available at: www.thejournal.ie/un-somalian-famine-is-over-but-action-stll-needed-347449-Feb2012/. (Accessed: 19 January 2017).

Tvedt, T. (2004) *The river Nile in the age of the British political ecology and the quest for economic power*. London: I.B. Tauris.

UN. (2001) 'Secretary-General asks United States geographers to work with him to tackle climate change problems, Environmental degradation and Sustainable development', Press Release SG/SM/7732, 1 March, Available at: www.un.org/News/Press/docs/2001/sgsm7732.doc.htm. (Accessed: 6 February 2014).

UN. (2011) 'UN project seeks to make use of Nile's resources to fend off poverty and hunger', UN, 27 October. Available at: https://news.un.org/en/story/2011/10/393202. (Accessed: 12 September 2016).

UN (2001). Secretary-General asks united states geographers To work with him to tackle climate change problems, environmental degradation and sustainable development, UN Press Release SG/SM/7732, 2 March. Available at: http://unis.unvienna.org/unis/en/pressrels/2001/sgsm7732.html. (Accessed: 12 September 2016).

UNDESA. (2013) 'Water for life decade: Water scarcity', Available at: www.un.org/waterforlifedecade/scarcity.shtml. (Accessed: 23 January 2014).

UNEP. (2006) 'Environment, sustainable development and the Nile River Basin', Education for Sustainable Development Innovations. MESA University Partnership. United Nations Environment Programme (UNEP): Nairobi, Kenya. Available at: https://docplayer.net/2630438-Education-for-sustainable-development-innovations-mesa-universities-partnership-environment-sustainable-development-and-the-nile-river-basin.html. (Accessed: 9 November 2017).

UNEP. (2012) *Reneable resources and conflict, toolkit and guidance for preventive and managing land and natural resources conflic*. United Nations Interagency Framework Team for Preventive Action. New York: United Nations Environment Programme (UNEP).

Urdal, H. (2005) 'People vs. Malthus: Population pressure, environmental degradation and armed conflict revisited', *Journal of Peace Research*, 42(4).

Van der Molen, I. and Hildering, A. (2005) 'Water: cause for conflict or co-operation?' *Journal on Science and World Affairs*, 1.

Veerhoeven, H. (2013) 'Why a 'water war' over the Nile River won't happen. Al Jazeera, 13 June. Available at: www.aljazeera.com/opinions/2013/6/13/why-a-water-war-over-the-nile-river-wont-happen. (Accessed: 27 July 2022).

Veilleux, J.C. (2013) 'The human security dimensions of dam development: The Grand Ethiopian Renaissance Dam', *Global Dialogue*, 15(2).

Verhoeven, H. (2013) 'China shifts power balance in the Nile River basin', China Dialogue, 4 July. Available at: www.chinadialogue.net/article/show/single/en/6178-China-shifts-power-balance-in-the-Nile-river-basin. (Accessed: 17 February 2015).

Water Policy Program. (2002) 'Survey of Nile System Pollution Sources,' Report No.64, September. United States Agency for International Development(USAID): Egypt.

Waslekar, S. (2013) *Blue peace for the Nile*. Mumbai: Strategic Foresight Group.

Waterbury, J. (1979) *Hydropolitics of the Nile Valley*. New York: Syracuse University Press.

Wheeler, K.G. *et al*. (2016). Cooperative filling approaches for the Grand Ethiopian Renaissance Dam, *Water International*, 41(4).

Wheeler, K.G. *et al.* (2020) 'Understanding and managing new risks on the Nile with the Grand Ethiopian Renaissance Dam,' *Nature Communications* 11. (5222)

Wiebe, K. (2001) 'The Nile River: Potential for conflict and cooperation in the face of water degradation', *Natural Resources Journal*, 41(3).

Wolf, A.T., and Newton, J.T. (2008) 'Case study of transboundary dispute resolution: the Nile waters Agreement', *Institute for Water and Watersheds*. Available at: www.transboundarywaters.orst.edu/research/case_studies/Nile_New.htm. (Accessed: 4 June 2014).

Wolf, A.T. (1996) 'Middle East water conflicts and directions for conflict resolution', Food, Agriculture, and the Environment Discussion Paper 12, March. Washington, D.C. International Food Policy Research Institute.

World Bank. (2008) 'Nile Basin initiative institutional strengthening project,' InfoShop – World Bank, Report No. AB3787. Available at: https://projects.worldbank.org/en/projects-operations/project-detail/P110616. (Accessed: 14 November 2016).

Xercavins, J. (1999) *Carrying capacity in East Sub-Saharan Africa: A Multilevel integrated assessment and a sustainable development approach*', inpublished thesis. Department of Mining Engineering and Natural Resources: Universitat Politècnica de Catalunya.

Yacob, Y. (2004) 'From UNDUGU to the Nile Basin Initiative: An enduring exercise in futility', Addis Tribune. Available at: http://allafrica.com/stories/200402200793.html. (Accessed: 24 October 2013).

Yacob, Y. (2014) 'Blue Nile sub-basin legal regime in historical context: Colonial treaties', Ethiomedia, 22 February. Available at: www.ethiomedia.com/14news/5701.html. (Accessed: 23 November 2016).

Yihdego, Z. (2013) 'The Blue Nile dam controversy in the eyes of international law', *Global Water Forum* .

5 The River Niger Basin

The River Niger Basin (RNB) is and has been important to the cultural and socioeconomic development of the West African region (Vincent *et al.*, 2002, p. 1). It is also an important lifeline for large parts of the Sahelian region (Wetlands International Africa, 2012). However, the basin, which is home to some of the poorest countries in the world, might be experiencing a new landscape of conflict with water resources being key players (Michel and Passarelli, 2014).

A combination of climate change, human population growth, and unsustainable resource use are threatening the RNB (Jagow-Schultz, 2017). This is in addition to hydropower dams and extensive irrigation schemes, which all affect the flow of the RNB (Wetlands International Africa, 2012). The unfolding scenario for water use in the RNB is one of increasing concern about access, equity, and the response to growing needs. Because the river is an important source of freshwater in this arid region, a coherent management agreement is needed between the riparian nations (Abebe, 2009, p. 123).

Description of the RNB

Located in Western Africa, the RNB starts in the highlands of Guinea, threading eastward mainly through Mali, Niger, and Nigeria before entering the Gulf of Guinea in the Atlantic Ocean. As the river nears the Sahara, where it forms an inland delta, its hydrology changes and it loses its flow and volume. The reason for the change is that the inland delta is an area of high evaporation composed of a number of slow-moving channels. Only after the Benue River does it become a large river once more (Abebe, 2009, p. 122).

The Niger is the third longest river in Africa after the Nile and Congo, flowing for 4,200 km. The drainage basin covers a surface area of just over 2.2 million km^2, extending into ten countries. These countries are Guinea, Côte d'Ivoire, Mali, Burkina Faso, Algeria, Benin, Niger, Chad, Cameroon, and Nigeria. The RNB has 76% of its area located in just three countries (e.g., Mali, Niger, and Nigeria) (*Figures 5.1* and 5.2) (Aich *et al.*, 2016, p. 3).

According to Mahé *et al.* (2009, pp. 7–8), from the standpoint of water resources, the RNB can diagrammatically be divided into four zones. Zone 1 is the

DOI: 10.4324/9781003366379-5

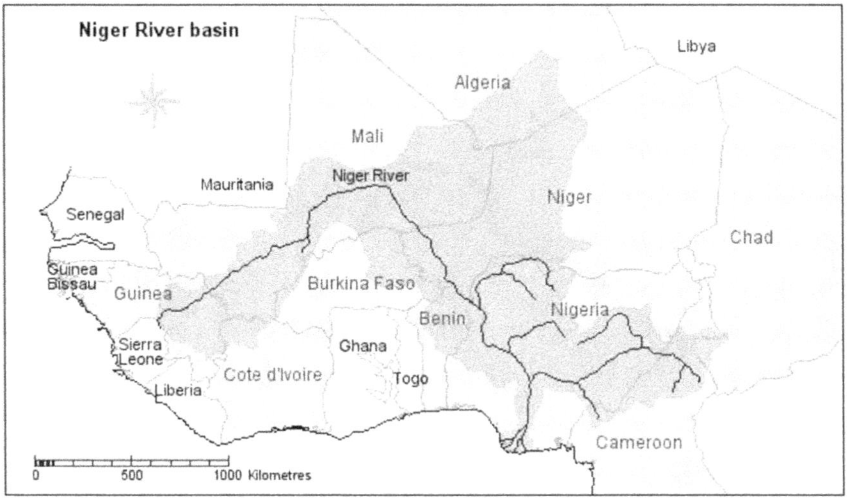

Figure 5.1 Map of the RNB.

Upper Niger Basin, which is found in Mali, Guinea, and Côte d'Ivoire. Zone 2 is the inland delta, which is situated entirely in Mali. Zone 3 is the Middle Niger Basin, which lies within Mali, Niger, and Benin. Zone 4 is the Lower Niger Basin, which lies between Cameroon, Nigeria, and Chad (Mahé *et al.*, 2009, pp. 7–8). The RNB's main tributary is the Benue River (Jagow-Schultz, 2017).

The RNB and its tributaries are an important water resource for multiple uses, including floodplain agriculture, irrigated agriculture, livestock wetland grazing and watering, production of hydropower, and domestic and industrial uses of water (Goulden and Few, 2011, p. 45). The RNB and its tributaries, with a population growing at approximately 3% a year on average (Andersen *et al.*, 2005, p. 1), are a key source of water for the drier regions within the western Sahel zone and a spinal column for the economy of the estimated 100 million people living in the basin (Goulden and Few, 2011, p. 5).

The dependency of riparian countries varies with location. Mali, Niger, and to a lesser extent Nigeria, compose large portions of the basin and rely heavily on the Niger for water. Benin, Cameroon, and Guinea compose only 14% of the basin land area but provide >80% of the water. Burkina Faso, Chad, and Côte d'Ivoire are the least dependent on the Niger for water but would benefit from the inclusive management of basin resources (Abebe, 2009, p. 122). The RNB is thus a key source of water, especially for the western regions of the Sahel (Goulden and Few, 2011, p. 16). Since the RNB is shared by ten countries, water resources management is coordinated at the international level by the Niger Basin Authority (NBA) (Goulden and Few, 2011, p. 21).

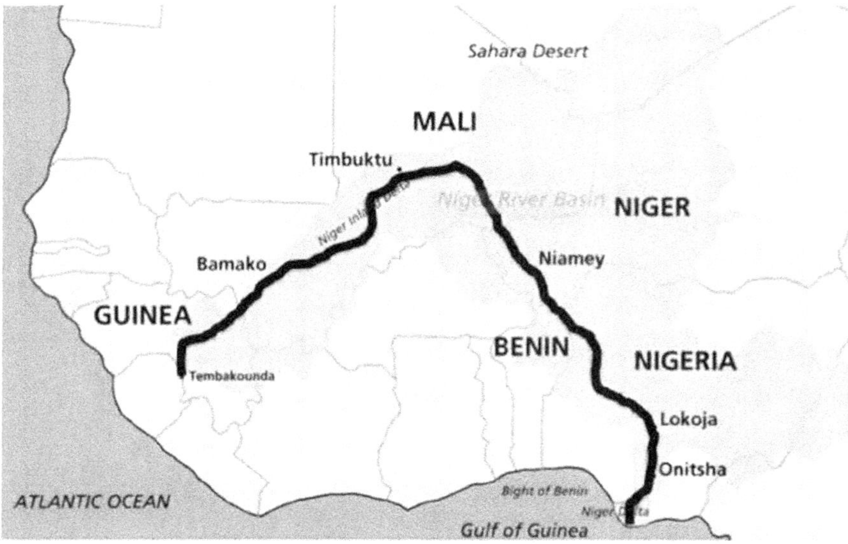

Figure 5.2 Map of the RNB.

An analysis of the environmental scarcity in the RNB

Population growth affecting water resources in the RNB

Population growth and rising resource consumption affect resource availability (Homer-Dixon, 1999a, p. 112). The RNB is inhabited by 120 million people across 11 countries (Blumstein, 2016), with a growth rate of approximately 3% per annum. Homer-Dixon (1999a, p. 52) argued that such an increase in population size and the increased per capita demand associated with it would undoubtedly increase the demand for land and water resources in the basin. It will simultaneously decrease supply by contributing to resource depletion and degradation (Homer-Dixon, 1999a, p. 52).

The population of Nigeria is 190 million people (Worldometers, 2017a), with 18 million in Mali (Worldometers, 2017b) and just over 20 million in Niger (Worldometers, 2017c). According to Olomoda (2002, p. 15), the majority of this population are farmers; however, migration to urban areas is rapidly taking over as a result of the unfavourable climate (e.g., poor rainfall and desert encroachment) and famine.

The rate of urbanisation has been rapid throughout most of West Africa since the 1950s. Several of the resulting urban agglomerations fall within the RNB, some of them located on the banks of the river, such as Niamey and Bamako. Access to improved water sources is an issue throughout most of the basin, and the projected population growth for the next few decades will increase the need

(United Nations Environment Programme (UNEP), 2010, p. 61). However, the growth of large cities along the river's banks has not been accompanied by the development of wastewater collection and treatment plants, whether for domestic or industrial wastewater (Andersen *et al.*, 2005, p. 56).

The population increase in countries, such as Niger, which gets about 90% of its water from the RNB, or countries like Nigeria and Mali which depend on the RNB for irrigation and farming, poses a serious demand and supply challenge. Steady increases in the populations of these areas are damaging the resources of the surrounding area (Ayuba, 2016). Not to mention that this area is one of the most impoverished in the world with all the countries, except for Cameroon, having >50% of their populations living below the poverty line. This makes it extremely difficult for development to occur in the region (World Hunger, 2015).

Mali and Niger depend on the RNB and its tributaries for the majority of their water resources. Niger is highly dependent on river flows from upstream countries. At the same time, Nigeria contributes a large amount of flow to the river from rainfall within Nigeria (Goulden and Few, 2011, p. 16). Michel and Passarelli (2014) predicted that in the next 15 years, the population across Mali, Niger, and Nigeria would grow by 75%, soaring to 337 million. Further, more than half of these people will be living in cities (Michel and Passarelli, 2014).

The exploding urban population growth, mostly in developing regions, means that certain challenges related to water will occur, which will affect the sustainability of human urban settlements. These include the lack of access to safe water and sanitation and increasing water-related disasters, such as floods and droughts. In cities, the provisions for water and sanitation have been the most pressing and painfully felt when lacking (United Nations Department of Economic and Social Affairs (UNDESA), 2015).

The RNB is at the centre of an important migration flow from north to south, due to economic differences and the demand for labour. The poorer and drier countries of Burkina Faso, Chad, Mali, and Niger export their labour to the coastal and wetter countries of Nigeria, Côte d'Ivoire, Benin, and Cameroon, largely to produce cash crops (e.g., coffee, cocoa, and bananas). To a lesser extent, there is also migration of the Bozo and Somono fishermen to the large reservoirs on the RNB and pastoralists, with their stock, to the inland delta in the dry season (Andersen *et al.*, 2005, p. 66).

Migration and large-scale demographic movements have, largely, developed in response to the river and the seasons, with herders moving their cattle and sedentary farmers relying on the bounty of the inland delta and flow-recession agriculture to raise their crops. With population growth that averages 3% in the RNB, pressure on the existing resources has increased dramatically in places, leading to resource conflicts (Andersen *et al.*, 2005, p. 66). Moreover, rural populations throughout the region have traditionally responded to changing resource availability by migrating (Michel and Passarelli, 2014). Homer-Dixon's environmental scarcity theory indicated that environmental scarcity or changing weather patterns would produce environmental refugees. These are forced to leave their home region due to sudden or long-term changes to their local environment, which compromise their wellbeing or secure livelihood.

However, Homer-Dixon's environmental scarcity theory's *ecological marginalisation* perspective does not regard migration because of environmental scarcity as a good adaptive strategy. This is because high population densities in specific areas as a result of migration, combined with a lack of knowledge and capital to protect local resources, will cause severe environmental damage, chronic poverty, and violent disputes (Homer-Dixon, 1999a, p. 15).

The demand for water for various purposes and uses has prompted riparian countries to develop the basin. According to Wetlands International Africa (2012), the critical water flow of the Niger is under ever more stress from dam infrastructure for hydropower and extensive irrigation schemes. Dams in the Upper Niger, in Guinea Conakry, and southern Mali withdraw large amounts of water destined for the Inner Niger Delta.

Today, <150 large dams out of 1,300 on the continent are in West Africa. The region has <2 large dams per 100,000 km^2, compared with 4.3 dams per 100,000 km^2 in Africa as a whole. The RNB has 27 major dams and > 5,000 small dams. Many projects, especially multipurpose ones, are being initiated.

In most cases, the projects are conceived at the national level, even though the purpose of the dam is essentially regional, and the construction of a structure has, by definition, consequences beyond its borders. These consequences include changes in the hydrologic regime of the river, the destruction of natural habitats, and the agricultural production and ecosystem services that they provide for livelihoods and food security (Ayuba, 2016, p. 14).

The lack of a shared vision and a sustainable development action plan (SDAP) had led to uncoordinated projects being developed, with the attendant negative environmental and socioeconomic impact on other riparian countries downstream. Some of these projects had to be stopped due to international outcry (Olomoda, 2002, p. 22). The development of new dams not only raised ecological concerns but also provoked difficult negotiations over equitably sharing the resources (Diarra, 2011). This problem, combined with a lack of shared visions, hinders development (Olomoda, 2002, p. 22).

Degradation and depletion of water resources in the RNB

According to Homer-Dixon (1999a, p. 48), because of a drop in the supply of water resources as a result of increased demand, it is assumed that the resource in question will shrink. Unfortunately, people's reliance on renewable resources has led to the depletion and degradation of these resources faster than they are being renewed.

During recent decades, civil war and unrest in several countries that border the RNB have made it the recipient of large numbers of people displaced by violent conflict, adding pressures and stresses to already fragile lands. For example, in 2003 >25,000 refugees settled in the Fouta Djallon and Mount Nimba regions of Guinea. This led to increased degradation of the highlands, with rapid deforestation and associated land degradation, soil erosion, gullying, and loss of productive lands. The loss of absorption capacity causes rapid runoffs, with the consequences of high sedimentation, the siltation of existing infrastructure, floods, and changes in river flows affecting downstream (Andersen *et al.*, 2005, p. 66).

The RNB is one of 11 African international river basins, more than half of which are approaching or are already under situations of water stress at the basin level (Goulden and Few, 2011, p. 6). According to the World Bank (2021), the Niger is considered to be one of the rivers most affected by freshwater shortages. This crisis has, therefore, led to competition among the Niger water users over already limited water resources.

Competition over scarce water resources is increasingly a source of tension in the RNB (Abebe, 2009, p. 123). Water use is increasing, resulting from agriculture, domestic uses, and hydropower (World Bank, 2021). This has led to the degradation of the environment because of low flows, sedimentation of river beds directly related to deforestation, over-farming of fragile soils, industrial and household pollution, and loss of arable and pasture lands (Abebe, 2009, p. 123).

Pollution caused by these factors is making the Niger's water increasingly unusable (World Bank, 2021; Simpkins, 2010). The degradation of water quality is a significant problem for the Niger (Andersen *et al.*, 2005, p. 56).

The depletion of the RNB because of unsustainable use also proves costly for riparian countries. For instance, the UNEP and the UNDP calculated the annual losses of Mali. Both these agencies estimated that, because of water scarcity, water-related illnesses, pollution, waste, deforestation, and desertification, Mali alone lost 20% of its gross domestic product (GDP). Similarly, the World Bank made its calculation on the annual costs to Nigeria because of environmental degradation. According to the World Bank, Nigeria's annual cost was estimated at from 5% to 10% of its GDP (Michel and Passarelli, 2014).

For the most part, cities along the river have not developed collection and treatment plants for either industrial or domestic wastewater. In addition to urban pollution sources, agricultural runoff, particularly fertilisers, has been found at several sites.

In the coastal delta, oil production has been the source of a host of environmental issues. Millions of barrels of oil have been spilt in the delta's oil-producing region (UNEP, 2010, p. 62). According to Phil Vernon, International Alert's director of programmes for Africa and peacebuilding issues

> You've got vulnerable people, vulnerable households, vulnerable communities living within… fragile systems, governance systems, [and] environmental systems,
>
> (Diamond, 2011)

The scarcity of water in the RNB has led to conflict between farmers and herders (Vincent *et al*, 2002, p. 1). Furthermore, Michel and Passarelli (2014) asserted that these environmental pressures and resource scarcity in the Niger region have played a part in producing widespread insecurity and militancy in the region. These have emerged because of a dearth of economic opportunities due to dwindling resources. They blame the 2013 radicalism of the Islamic terrorist group associated with Al Qaeda in Mali and the emergence of Boko Haram in neighbouring Nigeria on resource scarcity.

First, they asserted that in Mali, these radical Islamist factions were able to seize control of much of the nation's territory by exploiting a pre-existing revolt among the region's Tuareg peoples, first absorbing and then displacing the Tuareg insurgent forces. Crucially, this long-standing Tuareg rebellion first arose, in part, from the perceived indifference or inability of a distant central government to provide relief to Tuareg regions plagued by sustained drought in the 1970s and 1980s. Bamako eventually dampened this initial rebellion with formal peace negotiations and promises of development; however, assistance proved slow, and the north remained marginalised. Tuareg grievances continued and the insurgency ultimately returned (Michel and Passarelli, 2014).

In neighbouring Nigeria, many observers, such as the International Crisis Group, link the rise of Boko Haram to the decline of the once-dominant agricultural sector in Nigeria's impoverished northern states and popular frustrations at government failures to deliver water, energy, roads, and other services. With faltering agriculture and feeble infrastructure hobbling the northern economy, poverty and unemployment rendered the region's youth more vulnerable to Islamic radicalisation and Boko Haram recruitment.

Worse, the instability in Nigeria has now come full circle, threatening to deepen the cycle of insecurity as violence in the northern states further depresses agricultural production, disrupts harvests, and displaces farmers fleeing the fighting (Michel and Passarelli, 2014). Although these tensions and conflicts are localised, the capacity of water scarcity to intensify these conflicts beyond borders should not be underestimated.

Major environmental degradation in the RNB results from either natural or anthropogenic causes. Natural causes relate to climatic variability and change, in particular, the decrease in rainfall in the basin since the late 1970s. The major anthropogenic causes are land degradation, deforestation, and soil erosion that have taken place in the watershed, in large part because of increased demographic pressure (Andersen *et al.*, 2005, p. 67).

Land degradation, climate change, industrial and domestic issues, and migration are fundamental causes of the increase in the scarcity of land and water resources in the RNB (Homer-Dixon, 1999a, p. 9).

According to Andersen *et al.* (2005, p. 67), four principal environmental issues – land degradation, water degradation, deforestation, and biodiversity loss – have a synergistic effect on water resources in the basin. First, land degradation, in the form of erosion, is a result of inappropriate agricultural practices. These practices include bushfires, clearance for rice paddies, extensive cultivation, overgrazing, and the reduction in wetlands from drainage.

Water degradation, mainly the deterioration of water quality, is a result of the nonpoint source impacts of the pesticides and fertilisers used in agriculture, the point source urban pollution, and the lack of sanitation infrastructure (sewage) (Andersen *et al.*, 2005, p. 67).

Second, Jagow-Schultz (2017) saw habitat alterations as threatening to the wellbeing of the RNB ecosystem. They noted that these include dams, which affect the flow and sediment regimes of the river, in addition, to directly fragmenting and destroying aquatic habitats, irrigated floodplain agriculture, which displaces productive habitat, and the increasing discharge of sewage and other pollutants into the river.

Furthermore, in Nigeria petroleum development has contributed to several environmental problems, which pertain to water resources. In the delta, these environmental problems are usually caused by accidental pollution and illegal siphoning. In addition to petroleum development, the exploitation of coal, iron, gold, and other mineral resources is an environmental threat to the basin (Andersen *et al.*, 2005, p. 57).

This pollution of water resources in the basin has also led to significant environmental conflicts in Nigeria's oil-rich Niger Delta. Ethnic rebel groups, such as the Movement for the Emancipation of the Niger Delta, have long cited the chronic pollution of their communities' land and water among the grievances motivating their attacks on petroleum operations and guerrilla warfare against the Nigerian state (Michel and Passarelli, 2014). Although ethnic differences may be the key to these conflicts, resource scarcity that results from degradation is a multiplier of already existing tensions.

Third, at the moment the region relies mostly on its agricultural sectors to feed its expanding populations. Half of all water withdrawals in Nigeria, two-thirds in Niger, and 98% in Mali go to agriculture. Because of this dependence on the agricultural sector to feed the growing population, the water supplies are increasingly becoming stretched. The fact that drip irrigation schemes are not used in the basin, which could slash agricultural water demands when enhancing yields, exacerbates the situation. It is the lack of such irrigation systems that serve to exhaust water and land resources; therefore, exacerbating depletion, deforestation, and desertification (Michel and Passarelli, 2014).

Biodiversity loss is caused by habitat destruction and the subsequent increase of invasive species, which are in turn caused by inappropriate fishing practices, deforestation, and land conversion for agriculture (Andersen *et al.*, 2005, p. 67).

Fourth, Andersen *et al.* (2005, p. 67) indicated that the deforestation that occurred in the basin was the result of an increase in the need for energy and limited access to electricity. As a result, people in the basin use wood and charcoal for domestic purposes (which also contributes to land degradation).

Finally, the main impact of environmental change on water resources in the Niger is related to changes in river flow, pollution, and changes in the water supply. Climate change and variability predictions suggest an aggravation of the situation from a decrease in rainfall (World Bank, 2021). Goulden and Few (2011, p. 5) asserted that the RNB has experienced significant climate variability during the twentieth century, making it suitable to study the links between climate and conflict.

The effects of variations in climate are, and have long been, part of the reality of life in the RNB, mingled with other elements of the social and economic reality (Goulden and Few, 2011, p. 47). The climate in the RNB is marked by a high degree of variability.

Rainfall, the river flows, and temperatures already fluctuate a great deal and could become even more variable in the future (Diamond, 2011). Much of the basin is in a region of low rainfall, with high natural variability in rainfall and stream flows that affect land and water resources and people's livelihoods (Goulden and Few, 2011, pp. 5–6).

Climate variability is displayed in seasonal variations in rainfall over the basin and year-to-year variability in rainfall. In some years, this variability manifests as extreme events, which include droughts (recurrent in many parts of the basin) and severe floods. These have a major impact on lives, livelihoods, wellbeing, and the productivity of sectors (Goulden and Few, 2011, p. 45).

Drought and reduced water availability have forced rural communities, notably farming and cattle herding families, to migrate south to more humid conditions, increasing pressure on the remaining depleted floodplains, wetlands, and arable land (Abebe, 2009, p. 123). The ensuing intensity of competition has already caused localised conflict in semi-arid zones. There are signs of disputes based on ethnic dimensions as hardships are blamed by indigenous groups on *outsiders*, rather than affected communities seeing their difficulties as shared *conflicts of interest* (Abebe, 2009, p. 123).

The RNB was seriously affected by recurrent severe droughts, particularly during 1968–1973, 1982–19825, and 1990–1991, which caused great distress in the region (Amadou *et al.*, 2014, p. 3). The Sahel, the area just south of the Sahara, serves as the best example of the effects of severe droughts caused by climate change. The Sahel has experienced the worst effects of water scarcity, hotter temperatures, and longer dry seasons (UN, 2007, p. 18). For instance, in the 1970s and 80s, the Sahelian droughts spanned the RNB causing famine, forcing the dislocation of people, and destroying livelihoods. A period of reduced rainfall across the Sahel began in the early 1970s and continued to the 1990s, with two periods of very severe drought in the early 1970s and 1980s (UNEP, 2010, p. 62).

According to Jagow-Schultz (2017), the Niger dried up completely for several weeks in 1985 at Malanville in the Benin Republic. These developments may well be exacerbated by the lack of state and regional capacity to manage the effects of climate change (Werz and Conley, 2012, p. 3).

Nigerian President Mohammadu Buhari said the basin is experiencing significant degradation of its resources under the combined effects of climate and human activities. The president noted that the phenomenon is,

> heavily compromising the future of the ever-increasing populations of the basin through the intensity and number of environmental, social and economic problems it is inducing.

Buhari, who was represented by the Minister of Environment, Ms Amina Mohammed, said part of the processes for achieving a shared vision for the sustainable development of the basin led to the establishment of an SDAP for the basin (e.g., 2008–2012). The newly adopted 2016–2024 operational plan, with its climate resilience investment plan, is expected to consume USD 7.2 billion (*National Mirror*, 2016).

According to Michel and Passarelli (2014), climate change will inevitably make the environmental situation worse. They further asserted that in addition to tripling water demands among the riparian countries in the past 30 years, long-term precipitation trends show a 30% reduction in rainfall since the 1970s, which has affected the RNB's flow. It has fallen by some 20%–50%. These precipitation patterns are likely to grow even more variable, threatening deeper droughts and stronger floods (Michel and Passarelli, 2014). With the lack of infrastructure, such as irrigation systems to deliver reliable water supplies, this situation raises concerns for vulnerable populations across the region.

Climate change destabilises communities by adding uncertainty and stress, according to International Alert's Phil Vernon, and when a community is already vulnerable, the impact of that uncertainty and stress is amplified and the potential for conflict is greater. Moreover, that uncertainty has important implications for how people think about responding to climate change. As a result, in communities based in the Niger region, climate-induced vulnerability led to some kind of conflict (Diamond, 2011).

Changes in water quantity and quality owing to climate change will affect food availability, stability, access, and use, subsequently leading to increased vulnerability for poor rural farmers who live in arid and semi-arid tropical Africa (Abebe, 2009, p. 124). Research on the RNB found that the effects of climate change in the region were pervasive and that *latent conflict* between groups (disagreements and disputes over damages to farmland and restricted access to water, but not physical violence) was common (Diamond, 2011).

Throughout the basin, river flows and rainfall have been decreasing since the 1970s, creating tension between two of the major communities living in the basin (e.g., farmers and pastoralists). Pastoralists are forced to travel further to bring their herds to water, and farmers are expanding their cropland to feed growing populations, reducing the pathways available to herders and their livestock (Diamond, 2011). Nonetheless, herders and farmers suffer the effects of climate change. Climate change has driven herders and farmers further onto marginal lands (Michel and Passarelli, 2014).

Within the Sahel, drought conditions persist leading to human pressure on the available land and water, which often gives rise to conflicts, most especially within the inner delta (Olomoda, 2002, p. 15). This, in combination with the less predictable precipitation attributed to climate change, will lead to diminishing water levels and water scarcity (Wetlands International Africa, 2012). The climatic future of the RNB remains uncertain; however, climate change is expected to have a key influence on water resources and human security through its impact on climate variability and extremes (Goulden and Few, 2011, p. 16).

Mali, Niger, and Nigeria have all taken steps at the national level to implement planning for adaptation to climate change. Mali and Niger have produced national adaptation programmes of action (NAPAs), supported by the UNDP and the Global Environment Facility. This is a requirement for those classified as least developed countries that are signatories to the United Nations Framework Convention on Climate Change (UNFCCC) (Goulden and Few, 2011, p. 22).

Mali's NAPA was completed in 2007 and was led by the Ministry of Environment, supported by UNDP. Niger's NAPA was completed in 2006 and was led by the National Environmental Council for Sustainable Development, with funding provided by Global Environmental Facility. Nigeria, since it is not classified as a least developed country, was not required to produce a NAPA. Nonetheless, with the support of the UNDP and Canadian International Development Agency (CIDA), it began producing a similar document in 2010, which it called its national adaptation strategy and plan of action and aimed to finalise it by the end of 2011 (Goulden and Few, 2011, pp. 22–23).

According to the London School of Economic and Political Science (LSE) (2015), in 2011, these efforts resulted in the publication of the National Adaptation Strategy and Plan of Action on Climate Change for Nigeria. The strategy outlines responses to climate change in key areas, such as agriculture (crops and livestock), freshwater resources, coastal water resources and fisheries, forests, biodiversity, health and sanitation, human settlements and housing, energy, transportation and communications, industry and commerce, disaster, migration and security, livelihoods, vulnerable groups, and education.

However, the policy document did not gain official support. Instead, in 2012, the executive council approved the adoption of a National Climate Change Policy and Response Strategy. This strategy aimed to provide a framework for responding to climate change-induced challenges, such as increased flooding and rising sea levels. There are also plans to create a National Strategic Climate Change Trust Fund and develop a nationally appropriate mitigation action document. In 2013, a national policy on climate change was finally approved and adopted by the federal executive council. This formed the basis for any new climate change law (LSE, 2015).

Uneven distribution of water resources in the NRB

As water continues to become scarce and unevenly distributed, and when groups start to disturb river flows, water will become a contentious issue, especially in sub-Saharan Africa where there is high population growth (Abebe, 2009, p. 124). The analysis of the RNB, as far as Homer-Dixon's environmental scarcity theory is concerned, is very important as structural-induced scarcity (uneven distribution of water resources) does not exist in its pure form.

The case of the RNB will enable this book to illustrate the theory's idea that the interaction of demand and supply-induced scarcity can produce structural-induced scarcity. This will justify the notion stated by the theory that the three forms of scarcity are not mutually exclusive, and their interaction has the likelihood of instigating the occurrence of each other. Structural-induced scarcity in the Niger can occur because of resource capture, which will be influenced by the depletion of resources because of population growth and the unsustainable use of water resources.

When powerful groups within a society recognise that a key shared resource is becoming scarcer (due to supply and demand pressures), they will use their power to manipulate the regimes governing resource access in their favour, thereby accumulating more resources at the expense of other riparian countries (Homer-Dixon, 1999a, p. 15). Homer-Dixon (1999a, p. 74) further argued that greed and the fear of the disadvantage that rising scarcity might produce can be one motivation.

The existence of dams on the river, combined with climate variability, has caused misunderstandings between countries sharing the river in the past (Goulden and Few, 2011, p. 21). In West Africa, as the number of large dam projects increases, the level of countries' interdependency continues, and as the availability of water reduces, a fertile ground for misunderstandings and tension among countries is created. Examples include:

- Senegal and Mauritania in June 2000 (*IRIN News*, 2000).
- Burkina Faso and Ghana in 1998, when the increase in water use by Burkina Faso reduced the amount of water going to the Akosombo hydroelectric power station (Gao, 2010).
- Benin and Niger, over the sovereignty of Lete Island, the meeting place of nomad pastoralists from Niger who settle there seasonally, and sedentary farmers from Benin (*IRIN News*, 2015).
- Niger and Nigeria when Niger constructed dams that reduced the water flow to Nigeria, which has large dams for hydro agriculture and energy (*Sahel Standard*, 2015).
- Cameroon and Nigeria, where the migration of Lake Chad attracted Nigerian immigrants to the Cameroonian part of the lake (Abebe, 2009, p. 124; Youssef, 2016).

According to Diarra (2011), the new dams not only raised ecological concerns but also provoked difficult negotiations over equitable sharing. This is a result of the divergent interests of riparian countries. Moreover, ecologists fear that new dams will have damaging effects, such as an uneven distribution downstream, particularly for the more than a million people who live in the Inner Niger Delta in Mali. For instance, large-scale irrigation projects and hydroelectric dams have caused the water level in the Niger's fertile and ecologically diverse wetland to fall by >20 cm, shrinking the annually flooded area by 900 km^2.

According to Blumstein (2016), the construction of dams affects water levels, which ultimately has an impact on people depending on fisheries, rice production, or livestock. Furthermore, the decrease in water levels means that the navigation of the river would also be constrained since the water would no longer be deep enough for large boats (Diarra, 2011).

Blumstein (2016) argued that existing competition over natural resources, and local conflicts, could therefore grow because of the environmental change caused by natural and human activities. The reason is that environmental change may shift the balance of power between states, thus producing instabilities that could lead to war. Moreover, Blumstein (2016) argued that this could further destabilise a region that was already characterised by high political instability.

Diamond (2011) blamed the worsening tensions in the basin on poor policy decisions. This also calls for environmental laws and regulations in the region to be modified to address these environmental problems in the basin (World Bank, 2021). Therefore, regime-based interstate cooperation is a reality that needs an explanation.

Given the existence of the three forms of scarcity in the RNB, and their argued potential to lead to a nascent environmental conflict, it is important to analyse the formation and capacity of regimes in the basin to address environmental scarcity and thus avoid nascent environmental conflicts in the basin.

Water regime formation and capacity in the RNB

The catchment areas covering several countries, such as the RNB, require a binding agreement for cooperation among the riparian countries (Olomoda, 2002, p. 13). A common approach to river basin development has been to develop river basin master plans. This makes sense because such a comprehensive approach facilitates the elaboration of a broad and holistic development plan and provides an overall blueprint for potential and planned water resources development in the basin (Andersen *et al.*, 2005, p. 59).

The nine RNB countries that form part of the basin's Inter-Governmental Organisation[1] are among the poorest countries in the world. Four of the nine basin countries are among the bottom 20 countries on the World Development Indicators scale, while on the UNDP's Human Development Index, seven countries are rated among the bottom 20, with Niger having the lowest Human Development Index rating in the world.

The need for development and investment in the region is evident, and the RNB holds tremendous potential; this is the reason for cooperation (Andersen *et al.*, 2005, p. 58). Furthermore, according to the World Bank (2015), today, the RNB is one of the most fragile developing regions of the world, by any measure.

When dealing with a shared river basin and in the absence of an effective basin organisation, most countries will plan unilaterally, as witnessed earlier in the chapter. In some river basins, member states have rushed to unilaterally build infrastructure projects, seeking to acquire rights ahead of any neighbouring states doing the same. The NRB is a case in point. In the absence of a cooperative agenda to which countries have committed and that assigns benefits to each member state, the pursuit of unilateral development will most likely lead to lose-lose outcomes.

The potential consequences of these unilateral actions would include increased tension and insecurity between member states, and lost opportunities for regional cooperation and integration.

For many years, the trend in each of the RNB countries has been towards a unilateral development of the river's resources. From the position of each nation-state, this makes perfect sense, particularly given the lack of a strong regional river basin institution through which cooperative developments could be leveraged, promoted, and instituted (Andersen *et al.*, 2005, p. 63).

The World Bank (2021) argued that the Niger has a poor legal framework at regional and national levels. It is also characterised by the inadequate implementation of available regulatory instruments, such as supranational institutions, created by riparian countries and responsive to a country's ministry in charge of water resources (World Bank, 2021).

Efforts towards basin-level cooperation in the RNB started with the 1963 Niamey Agreement (Act de Niamey), which defined the conditions of navigation on the RNB. The agreement also called for economic cooperation between basin states. A year later, the NRC – or Commission du Fleuve Niger/CFN – was established to promote and coordinate studies on the RNB's resources with a view to their future development. In 1980, the NBA was created with a mandate to speed up interstate cooperation in developing the basin's resources. The mandate embraced water development, energy, transport and navigation, agriculture, fisheries, livestock production, forestry, and industry. Faced with a situation of lethargy and unable to forward its ambitious mandate, the NBA saw its convention revised in 1968, 1978, 1979, 1980, and 1987 (Niasse, n.d).

1963 Niamey Agreement concerning the RNB Commission and the navigation and transport on the River Niger

The basin faces many environmental problems as evidenced previously. However, the RNB has the potential for a spectrum of development uses including domestic and industrial water supply, flood control, hydropower generation, irrigation, fisheries, navigation, and tourism (UN, 1984, p. 209).

Just after independence in the early 1960s and inspired by the Tennessee Valley Authority, the RNB riparian countries came together and developed a binding agreement to form the RNB Commission (NRC). The NRC was initially formed to foster cooperation in the field of management and use of the water resources of the Niger and to promote sustainable development (Olomoda, 2002, p. 13).

The convention during which these countries established the NRC took place in Niamey on 26 October 26 1963. A year later, on 25 November 1964 after many rounds of negotiations, an agreement concerning the RNC and the navigation and transport on the river Niger, also referred to as the Niamey Treaty was adopted (UN, 1967, p. 23).

Article 12 of the agreement relating to the agricultural and industrial use and development of the RNB clearly states its aims with regard to the development, use, and sustainability of the RNB. According to Article 12:

> They undertake further to abstain from carrying out on the portion of the River, its tributaries and sub-tributaries subject to their jurisdiction any works likely to pollute the waters, or any modification likely to affect biological characteristics of its fauna and flora, without adequate notice to, and prior consultation with, the Commission
>
> (UN, 1964, p. 27)

This agreement led to the establishment of the NRC to manage and develop the RNB as set out in Article 1 of the agreement. According to Article 1:

> There shall be established an Inter-Governmental Organization … which shall be called River Niger Commission
>
> (UN, 1967, p. 23)

However, the NRC's functionality was poor. As a result, three conventions, in 1968, 1973, and 1979 were established. At each convention, the following statements about the objectives were issued. First, the 1968 convention aimed at forging close cooperation between the member states for the sound development of basin-wide resources. Second, the 1973 convention emphasised the freedom of navigation on the river Niger among member states. Finally, the 1979 convention aimed at equal access to the basin's water resources for all users (Olomoda, 2002, p. 13).

1980 convention creating the Niger Basin Authority (with protocol relating to the Development Fund of the Niger Basin)

In 1979, the 1964 Niamey Agreement was revised. This led to the adoption of the 1980 Convention creating the Niger Basin Authority (with protocols relating to the Development Fund of the Niger Basin). This meant that the NRC had been replaced by the NBA (from a commission to an authority) to provide an institutional and legal framework to attract prospective donors and regain the confidence of its member states (UN, 1984, p. 209).

According to Article 21 of the 1980 convention:

> This Convention revises the Niamey Agreement, signed in Niamey on the 25th of November 1964, revised in Niamey on the 2nd of February 1968 and on the 15th June 1973, and in Lagos on the 26th January 1979
>
> (UN – Treaty Series, 1984, p. 216)

Although it revised the Niamey Agreement, the 1980 convention committed itself to inherit the Niamey Agreements' duties and obligations. For instance, Article 1(3) of the 1980 convention states that:

> The Authority inherits all the assets and assumes all the obligations of the River Niger Commission
>
> (UN, 1984, p. 209)

After the adoption of the Niamey convention and the creation of the NRC on 25 October 1964, the nine NBA member countries revised the NRC convention several times to improve its institutional performance. When these revisions showed no result, the NBA was formed to replace the NRC (Olomoda, 2002, p. 17). The NBA is still headquartered in Niamey, Niger.

The preamble of the 1980 convention states the reasons for the NRC being transformed into the NBA. According to the preamble, the convention saw the need to further promote the socioeconomic progress and development of member states. It also aimed to reaffirm the willingness for unity and solidarity in the organisation for the overall development of the RNB. The Summit of Heads of State and Government, which met on 26 January 1979 in Lagos, therefore, decided to transform the NRC into the NBA (UN, 1984, p. 209).

The institutional and legal framework to carry out its mandate went through four phases between 1980 and 2001. Phase I (1980–1986) saw the establishment of the Executive Secretariat. The most important assignment of this phase was to ensure that the aims and objectives set out in the 1980 convention were achieved. This was possible through financial support from donors. However, problems such as institutional weakness and inadequate finance persisted. As a result, the NBA could not carry out its assignment (Ayibotele, 2008, p. 1). Moreover, the NBA did not meet the aspirations of the member states. Some of these states signified their intention to change their membership to an observer status while others merely attended the meetings with scepticism (Olomoda, 2002, p. 13).

Phase II (1987–1993) was to adopt a revised 1980 convention to cure the policy shortcomings of the 1980 convention, restructure the authority and adopt a 5-year development plan. As such, this phase saw the reformation of the authority to focus on development and to strengthen it institutionally and financially. In addition, a revised convention was signed in 1987, which committed member states to acquaint themselves with development plans relating to aspects affecting the RNB.

Furthermore, member states were prohibited from carrying out any works on the section, tributaries, and subtributaries of the river that were likely to pollute the basin. However, the NBA failed to address the problems experienced in the first phase due to the notable lack of funds that confronted this phase and the attendant challenge of performing its duties of compiling essential data for long-term policy development (Ayibotele, 2008, p. 1)

In phase III (1994–1997), the problems experienced in phases I and II became worse, leaving the authority without political and policy direction. This led to the Economic Commission for African States stepping in to assist. Finally, phase IV (1998–2001) saw a resurgence of the authority. Most importantly, in this phase, the authority regained the confidence of its member states and its development partners (Ayibotele, 2008, pp. 1–4).

The NBA is mandated to foster, promote, and coordinate studies and programmes relating to the basin. It is further required to promote cooperation among member countries and ensure the integrated development of resources, notably energy, water, agriculture, forestry, transport, and communication. The NBA has worked to create an integrated development plan for the basin, focusing on cross-boundary projects (Abebe, 2009, p. 125).

The Niger riparian countries had realised that all the NBA's goals were possible if well managed. These are imperative aspects of development, as indicated in Article 3 of the 1980 Convention creating the NBA (with protocol relating to the Development Fund of the Niger Basin), which states,

> the aim of the Authority is to promote the cooperation among Member States and to ensure an integrated development of the Niger Basin in all fields, by developing its resources, particularly in the fields of energy, water resources, agriculture, animal husbandry, fishing and fisheries, forestry and forestry exploitation, transport, communications and industry,
>
> (UN, 1984, p. 209)

One of the NBA's major initiatives was the adoption and implementation of an SDAP, which most significantly involved managing several hydroelectric and agricultural dams built along the river. Therefore, if one riparian country wanted to build a dam, it would have to take the plans to the NBA for a decision. Representatives from each of the RNB countries formed part of the NBA; therefore, each country had a chance to comment on water resource development plans (Goulden and Few, 2011, p. 21).

In Abuja in February 2002, the RNB heads of state agreed to develop a management framework for the basin by preparing a shared vision process (Andersen *et al.*, 2005, p. 59). The shared vision process was completed in 2008 (Talbi, 2013). It laid out cooperative actions that defined needs and priorities and identified management, development, and investment actions (Andersen *et al.*, 2005, p. 59).

The shared vision was an expression of the countries' commitment to promoting a framework for enhancing cooperation and sharing benefits derived from the RNB's resources. The shared vision process encompassed several objectives. The first objective was political; to formulate a statement on the sustainable development of the RNB to be adopted by the RNB heads of state. Such a statement would include commitments to, and goals for, cooperation that would lead to joint developments in the basin. The second objective was operational; to prepare the SDAP for the basin. The SDAP was seen as an appropriate instrument to realise the countries' commitment to addressing the challenges of the basin. It included an innovative planning and priority-setting approach to defining the development opportunities in which member countries could jointly participate. The shared vision's third objective was financial; to mobilise resources from member countries and international donor partners to implement the SDAP (Andersen *et al.*, 2005, p. 59).

The shared vision process provided an opportunity for the nine member countries to reaffirm at the highest level of authority their willingness to collaborate and to place a higher emphasis on new values, such as equitable access, sustainable development, and public participation (Niasse, n.d.).

According to Talbi (2013), noticeable achievements because of the shared vision process include:

- A 20-year SDAP for the basin was approved in 2007 by the Council of Ministers of the NBA.
- A 20-year investment programme for the basin and the Water Charter was approved in 2008 by the Heads of State and Government Summit.
- The Water Charter was signed in 2009 and ratified in 2010. In addition, Annexures to the Water Charter were prepared, and the first was adopted and signed in 2012.

The NBA derived funds from member states and international donors and participated in cooperative projects with organisations such as the World Wildlife Fund, Wetlands International, and foreign donor governments (Abebe, 2009, p. 125). Nevertheless, President Buhari reiterated that the NBA faced the challenge of financial capacity as member states failed to fulfil their financial obligation (*National Mirror*, 2016). This was a challenging factor confronting the organisations' capacity to execute imperative tasks relating to the development and management of the basin. This was also a major reason that environmental issues persisted unabated.

The environmental challenge also persisted despite a provision in Article 4 of the convention. Among other issues relating to water control and use, Article 4 further committed the convention to regulate the main waterway flows and drainage, controlling floods, preventing, and controlling drought and desertification, and preventing soil erosion and sedimentation (UN, 1984, p. 211).

That the challenges persisted is indicative of poor policy execution and implementation by member states. Furthermore, it signalled the lack of political will by member states and exposed the NBA's lack of capacity to hold member states (signatories) accountable for actions that were hazardous to the RNB, and which were committed in their respective countries.

1987 revised Convention for the establishment of the NBA

On 27 October 1987, the revised convention for restructuring the NBA was adopted with three objectives in mind. First, it aimed at harmonising and coordinating national development policies on basin-wide resources. Second, the NBA's objective was to formulate a policy and implement an integrated plan to develop the basin. The NBA's third objective was to design, build, and maintain common structures and projects (Olomoda, 2002, p. 17; International Environmental Agreements (IEA, 2016).

These objectives are all important because to effectively and cooperatively counter the threat posed by environmental issues in transboundary river basins, regime member countries will have to harmonise national laws, implement integrated plans, and establish common structures to effectively implement transboundary agreements. Institutions can range from independent national bodies that coordinate policies, to joint institutions that formally receive decision-making power from the respective national governments (Brels *et al.*, 2008, p. 13). Harmonisation is normally necessary as a means of avoiding some type of dispute as a result of water scarcity (Earle, 2005, p. 56).

Despite these new objectives, the organisation continued to face several operational and management difficulties. Some of them are a result of inadequate political will to back the mandates and of various economic, social, and political problems faced by member states. Similarly, since the degree to which the countries share in the basin varies, with Nigeria having 30% and Chad at 1%, the riparian countries have different levels of interest and commitment in developing and managing the basin. However, in an attempt to find a balance with regard to the equitable and accommodative financial accountability of developing and managing the basin, a new sharing formula was proposed (Olomoda, 2002, p. 17).

The Sixth Summit of Heads of State and Governments of the NBA member countries finally adopted, in December 2000, a new formula that allowed for dues to be shared more equally than in the past. The new formula gives the three greatest beneficiaries of the RNB's resources (e.g., Mali, Niger, and Nigeria) a substantial increase in their shares, while the other member states receive corresponding budgetary relief. The new sharing formula ranges from 30% for Nigeria to 1% for Chad, which reduces the financial burden on the poorer nations (Olomoda, 2002, p. 20).

In addition to creating regimes in the RNB, Niasse (n.d.) argued that numerous agreements were established between two or more states in subsections of the basin. Examples include the 1990 agreement between Nigeria and Niger on the equitable sharing, conservation, and development of their common water resources; the 1998 agreement between Niger and Mali relating to cooperation on the use of water resources of the Niger; the hydroecological management project of the Upper Niger between Guinea and Mali; and the 2000 Nigeria–Cameroon Protocol Agreement aimed at coordinating the release of water from dams.

These agreements came after the establishment of the NBA. They highlight the relevance of the subsidiarity principle more than they challenge that of basin organisations. Furthermore, it is important to emphasise that in at least one case (Nigeria–Cameroon agreement), the NBA facilitated discussions between both countries concerned (Niasse, n.d.).

However, Olomoda (2002, p. 22) maintained that despite several adoptions of agreements and the transformation of the NRC into the NBA, several major institutional problems persist. These include:

- A lack of clearly defined objectives and suitable regional strategies based on a clear shared vision or master plan acceptable to riparian countries for the development of the basin.
- A lack of legal and institutional mechanisms for benefit sharing to prevent and manage water-related conflicts.
- An inadequate framework for defining institutional objectives and policies.
- Inadequate institutional and operational capacity to plan, design and execute projects.
- A lack of community involvement in the projects, from grassroots project planning to project execution.
- A weak institutional capacity and vision for planning and coordinating the implemented action.
- The lack of cooperation between riparian member countries.
- An insufficient mechanism for consultation among stockholders.
- A lack of adequate knowledge of national policy, activities, and orientation of water resources projects at a national level.
- A lack of cooperation and adequate consultation between institutions, intergovernmental organisations, and donor agencies.

Homer-Dixon (1999a, p. 4) asserted that environmental scarcity will leave political institutions fragile due to them having less, if any, financial, material, and human capital (Homer-Dixon, 1999a, pp. 4–5). Although the loss of water supplies physically manifests as local and regional problems, water scarcity by itself does not trigger war. The main issue that determines the emergence, deterrence, or both of war is the governance of water. This governance failure and poor coping mechanisms can contribute to unrest and conflict (Abebe, 2009, p. 124).

Conclusion

Climate change, unsustainable water use, and booming population growth in Mali, Niger, and Nigeria are driving a looming resource shortage, exacerbated by strained infrastructure and regime capacity, that risks pushing them past breaking point (Michel and Passarelli, 2014). Due to the uncertainties created by climate change for vulnerable communities in the basin, development and adaptation policies must be flexible enough to cope with extreme variability, in the wet and dry conditions of the RNB.

Boosting resilience will be key to ensuring that vulnerable communities have the flexibility they need to respond to future crises. It is evident from this chapter that policymakers are left with the urgent crisis of climate change. A solution that is inherently time-intensive (i.e., building resilience in vulnerable communities down to the individual level) is important simply because individuals in the region experience and react to climate change differently.

Economic deprivation because of environmental degradation may have already taken its toll, contributing to destabilising much of the region and potentially leading to a nascent environmental conflict, which could threaten global security. These countries are faced with pressing challenges in the form of socioeconomic change, resource stress, and violent extremism. The challenge facing the countries of the RNB is to find ways in which the river's development potential can be realised. It is undeniable that the economic status of the region, being home to some of the poorest countries in the world, affects the formation and effectiveness of environmental regimes.

According to Niasse (n.d.), efforts towards promoting transboundary cooperation were constrained for several decades by factors, such as the size of the basin, the high number of basin countries, divergent interests among basin countries, and financial issues. However, they further state that in the past ten years, and especially since 2000, the NBA has gained vitality and has become more credible in the eyes of member states.

The main problem facing the NBA, which regulates all basin activity, is that it lacks political support from the member countries. As a result of this lack of political support, there is a greater lack of technical and financial support. For example, only the Federal Republic of Nigeria and the Republic of Benin have paid all their contributions to the NBA. The NBA also shows institutional weakness, specifically in an inability to attract the support of donors or the involvement of member countries in projects.

Each of the environmental issues mentioned above is extremely detrimental to the peace and security of the RNB. The NBA can play an important role in increasing awareness of the transboundary impact of socioeconomic pressures on natural resources. In addition, the NBA can assist member countries to identify investment resources to protect the basin and its watershed. This would provide benefits, not only to the country where the protection is undertaken but also to the downstream countries, which will benefit from the protection of the source through improved water quality, controlled run-off, and reduced sedimentation (Andersen *et al.*, 2005, p. 67).

In some cases, the downstream benefits from watershed rehabilitation and protection investments upstream might be so significant that the downstream countries receiving the benefits will invest jointly in such activities. The subsidiarity principle will help the NBA to identify areas, as part of the SDAP, where the institution will have a comparative advantage over well-established national and local agencies that are also charged with working on these matters. The key questions are whether the action in this field is part of the core NBA mandate and whether a transnational approach by the river basin organisation is the most appropriate way to tackle a particular issue (Andersen *et al.*, 2005, p. 67).

The lack of adequate solutions aimed at addressing environmental scarcity, coupled with increasing environmental degradation, are among the causes of the present alarming growth of poverty in the region. The countries should seriously carry out policies relating to the shared vision for the integrated development and management of projects. This would ensure benefit-sharing and enhance the rapid socioeconomic development of the basin (Olomoda, 2002, p. 22).

This book found that the RNB faces serious environmental challenges, such as depletion and degradation of the basin and land degradation caused by population growth industries, domestic use, and climate change. In addition, there is a serious challenge of unequal access to, distribution of, or both for the RNB. Theoretically, it is viable to argue that these environmental challenges and scarcities are placing pressure on the four member states to meet domestic demands. The theory defined environmental scarcity in terms of the three main sources of environmental stress that together form environmental scarcity, for example, population growth, environmental change, and unequal distribution of resources (Homer-Dixon 1994, pp. 8–11).

These three forms of scarcity are central to the emergence of environmental conflicts. The increasing emergence of these environmental challenges and the threat of nascent environmental conflicts in the basin further question the capacity of regional and basin-wide regimes governing the basin. Moreover, Homer-Dixon's environmental scarcity theory emphasised environmental scarcity (e.g., water and land, which are the most important environmental resources as environmental causes of violent conflict) (Homer-Dixon 1994, p. 18), caused by population growth, resource maldistribution and resource depletion or degradation, including the interaction among these three sources of scarcity (Solomon, 2014, p. 74).

Homer-Dixon's environmental scarcity theory centres around three arguments regarding these forms of scarcity and their relationship to violent conflicts (e.g., demand, supply, and structural-induced scarcity). First, according to the theory, population growth and large population movements caused by environmental stress would intensify pressure on shared resources; therefore, limiting the ability of states to meet local demands. This will eventually provoke conflict. Second, the theory argues that the decreasing quality and quantity of land and water resources would make conflicts between riparian countries probable. Finally, the theory argues that the unequal distribution of resource scarcity would increase economic deprivation and disrupt key social institutions, also making conflict probable (Solomon, 2014, p. 74).

The book further found that little attention is given to the threats emanating from climate change and ozone depletion, which are central to affecting the natural cycle of resource production, as far as policymaking is concerned.

With the use of Homer-Dixon's environmental scarcity theory, a nascent conflict can occur in three forms. First, conflict may arise because of competing interests between different water users and (unfair) water allocation. Second, tensions may arise between upstream and downstream populations regarding water pollution by human activities (industry and agricultural contamination of water resources, ecological pollution due to untreated wastewater and dumping). Third, possible tensions and conflict may arise due to climate change (droughts and floods), which may affect food security and trigger conflicts in future.

It has been made evident that a link exists between the functioning, effectiveness, or both regimes and environmental stresses when generating conflict. Van Schaik and Dinnissen (2014, p. 33) made a substantial argument outlining the link between the effectiveness of regimes and environmental stress. They argued that environmental stress coupled with weak regimes lay a strong foundation for the emergence of interstate conflicts.

Note

1 While a small area of Algeria falls within the Niger Basin, it is not a member of the NBA since it is neither a significant contributor nor a major user of basin water (McKinney, 2012)

Bibliography

Abebe, L. (2009) 'Natural resource conflicts in West Africa: The case of the RNB', in Tadese, D.W. (ed.) *Climate change and transboundary water resource conflict in Africa.* Mombasa: Institute for Security Studies Workshop report.

Aich, V. *et al.* (2016) 'Time series analysis of floods across the RNB', *Water*, 8(165).

Amadou, A. *et al.* (2014) 'Changes to flow regime on the RNB at Koulikoro under a changing climate', *Hydrological Sciences Journal*, 60(10).

Andersen, I. *et al.* (2005) *The RNB: A vision for sustainable management*. Washington, DC: The International Bank for Reconstruction and Development

Ayibotele, N.B. (2008) Integrated Water Resources Management (IWRM) in the Niger River Basin Case #46. Niger Basin Authority and Global Water Partnership. Available at: www.gwp.org/globalassets/global/toolbox/case-studies/africa/transboundary-organ ization-in-the-niger-river-basin-46.pdf. (Accessed: 1 February 2017).

Ayuba, H. (2016) 'Environmental risks and developmental challenges confronting river Niger and its basin', *8th African Organization of Supreme Audit Institutions Working Group on Environmental Auditing (Afrosai-WGEA) annual meeting*, Abuja, Nigeria. 12–16 September.

Blumstein, S. (2016) 'Calming the waters: Why we need to better integrate climate and water policy', *New Security Beat*, 23 August. Available at: www.newsecurityb eat. org/2016/08/calming-waters-integrate-climate-water-policy/. (Accessed: 1 March 2017).

Brels, S., Coates, D., and Loures, F. (2008) 'Transboundary water resources management: the role of international watercourse agreements in implementation of the CBD', *Secretariat of the CBD Technical Series*, 40.

Diamond, K. (2011) 'Climate change, water, and conflict in the Niger River Basin,' *New Security Beat*, 13 December. Available at:www.newsecuritybeat.org/2011/12/ climate-change-uncertainty-and-conflict-in-the-niger-river-basin/. (Accessed: 20 February 2017).

Diarra, S.T. (2011) 'West Africa: Niger River under pressure from dams', Inter Press Service (IPS) News Agency, 31 October. Available at: www.ipsnews.net/2011/10/west-africa-niger-river-under-pressure-from-dams/. (Accessed: 23 February 2017).

Earle, A. (2005) 'Hydropolitics in Southern Africa: What is the prognosis for peaceful development of shared watercourses?', Forschungsinstitut Wasser und Umwelt (FWU)/ University of Siegen Research Institute for Water and Environment FWU, Vol.3, Topics of integrated Watershed Management – Proceedings, Summer School. University of Siegen Research Institute for Water and Environment: Siegen.

Gao, Y. (2010) 'Transboundary water governance in the Volta River Basin', Available at: https://wikis.uit.tufts.edu/confluence/display/aquapedia/Transboundary+Water+Governance+in+the+Volta+River+Basin(Accessed: 23 February 2017).

Goulden, M., and Few, R. (2011) 'Climate change, water and conflict in the Niger River Basin,' USAID report, December . Norwich: International Alert and University of East Anglia.

Homer-Dixon, T.F. (1994) 'Environmental scarcities and violent conflict: Evidence from cases', *International Security*, 19(1).

Homer-Dixon, T.F. (1999a.) *Environment, scarcity, and violence*. New Jersey: Princeton University Press.

IEA. (2016) Available at: https://iea.uoregon.edu/treaty-text/1980-nigerbasinauthorityen txt. (Accessed: 23 February 2017).

IEA. (2016) 'Revised convention creating the Niger Basin Authority, International Environmental Agreement (IEA) database project 2002-2020. University of Oregon Environmental Studies Program: Eugene, Oregon', Available at: https://iea.uoregon.edu/treaty-text/3025. (Accessed: 23 February 2017).

IRIN News. (2000) 'Focus on renewed tension with Mauritania', 8 June, Available at: www.irinnews.org/feature/2000/06/08/irin-focus-renewed-tension-mauritania. (Accessed: 23 February 2017).

IRIN News. (2015) 'International Court rules that main disputed island belongs to Niger not Benin', 12 July. Available at: www.irinnews.org/report/55426/benin-niger-international-court-rules-main-disputed-island-belongs-niger-not-benin. (Accessed: 23 February 2017).

Jagow-schultz, T. (2017) 'Niger: Boomerang River in Sub-Saharan Africa', Worldwide Fund for Nature. Available at: http://wwf.panda.org/about_our_earth/ about_freshwater/rivers/niger/. (Accessed: 01 February 2017).

LSE. (2015) 'The global climate legislation study: Nigeria'. *Grantham Research Institute on Climate Change and the Environment*, 7 December. Available at: www.lse.ac.uk/GranthamInstitute/legislation/ countries/nigeria/. (Accessed: 23 February 2017).

Mahé, G., Liénou, G., and Adeaga, O. (2009) BFP Niger: Water availability and access. Montpellier and Colombo: IRD and CPWF, Research Institute in Montpellier, France.

Michel, D., and Passarelli, R. (2014) 'The Climate Wars are already here', *Foreign Policy*, 17 December. Available at: http://foreignpolicy.com/2014/12/17niger-river-basin-climate-wars-are-already-here/. (Accessed: 03 February 2017).

National Mirror. (2016) 'Stakeholders in fresh move to tackle problems of Niger Basin Authority', 31 May. Available at: http://nationalmirroronline.net/new/stakeholders-in-fresh-move-to-tackle-problems-of-niger-basin-authority/. (Accessed: 21 February 2017).

Niasse, M. (n.d.) 'The Niger Basin Authority. SHARE Toolkit: Case Studies', *IUCN Water Programme*. Available at: www.iucn.org/sites/dev/files/import/downloads/niger.pdf. (Accessed: 23 February 2017).

Olomoda I. (2002) 'Integrated water resources management: The Niger Basin Authority's Experience', in Castelein, S. (ed.) *From Conflict to Co-operation in International Water Resources Management: Challenges and Opportunities*, pp. 13-25.

Sahel Standard. (2015) 'War looms between Nigeria, Niger over River Niger', 4 May. Available at: http://sahelstandard.com/index.php/2015/05/04/war-looms-between-nigeria-niger-over-river-niger/ (Accessed: 23 February 2017).

Simpkins, G. (2010) 'Africa faces water war threats', *Africa Rising 21st Century*, 26 February. Available at: http://africarising2010.blogspot.co.za/2010/02/africa-faces-water-war-threats.html. (Accessed: 11 March 2016).
Solomon, H. (2014) 'Potential for cooperation rather than conflict in the face of water degradation: The cases of the Nile River and Okavango River Basins', *Southern Journal for Contemporary History* 39(1).
Talbi, A. (2013) '1 River Basin, 9 Countries, 1 Vision', The Water Blog – World Bank Blogs, 20 March. Available at: http://blogs.worldbank.org/water/1-river-basin-9-countries-1-vision. (Accessed: 23 February 2017).
UN. (1967) '1964 Agreement concerning the RNB Commission and the navigation and transport on the river Niger', Treaty series, 587–3(8507).
UN. (1984) '1980 Convention creating the Niger Basin Authority (with protocol relating to the Development Fund of the Niger Basin)', United Nations Treaty series, Vol.1346(22675). New York: UN.
UN. (2007) *Climate change: Impacts, vulnerabilities and adaptation in developing countries.* United Nations Framework Convention on Climate Change.
UNDESA. (2015) 'We're finally at the end of the UN Decade for Water 2005-2015 – It is time to say good-bye, United Nations Department Of Economic and Social Affairs, International decade for action water for life 2005–2015. United Nations. Available at: www.un.org/waterforlifedecade/water_cities.shtml. (Accessed: 1 March 2016).
UNEP. (2010) *Africa water atlas, division of early warning and assessment (DEWA)*. Nairobi: United Nations Environment Programme.
Van Schaik, L., and Dinnissen, R. (2014) 'Terra Incognita: Land degradation as underestimated threat amplifier', Clingendael: Wassenaar: Netherlands Institute of International Relations.
Vincent, J ., Aden, J ., Adriani, M ., and Dore, G. (2002) 'Africa region: Reversing land and water degradation trends in the Niger River Basin,' The United Nations Food and Agriculture Organization. New York: UNDP.
Werz, M., and Conley, L. (2012) 'Climate change, migration, and conflict addressing complex crisis scenarios in the 21st Century', Washington, D.C: Center for American Progress, Heinrich Böll Foundation.
Wetlands International Africa. (2012) *The Niger River and its deltas*. Available at: http://africa.wetlands.org/Whatwedo/NigerRiverandDeltas/tabid/2944/language/en-GB/Default.aspx. (Accessed: 1 February 2017).
World Bank. (2015) 'In the Niger Basin, countries collaborate on hydropower, irrigation and improved water resource management', The World Bank, 4 March. Available at: www.worldbank.org/en/news/feature/2015/03/04/niger-basin-countries-collaborate-on-hydropower-irrigation-and-improved-water-resource-management. (Accessed: 23 February 2017).
World Bank. (2021) 'World bank invests to combat water scarcity and food insecurity in Niger', The World Bank Press Release, 30 September. Available at: www.worldbank.org/en/news/press-release/2021/09/30/world-bank-invests-to-combat-water-scarcity-and-food-insecurity-in-niger. (Accessed: 26 January 2022).
World Hunger. (2016) '2016 World hunger and poverty facts and statistics, World Hunger Education Service'. Available at: www.worldhunger.org/2015-world-hunger-and-poverty-facts-and-statistics/. (Accessed: 1 March 2017).
Worldometers. (2017a) 'Nigeria population'. Available at: www.worldometers.info/world-population/nigeria-population/. (Accessed: 1 March 2017).

Worldometers. (2017b) 'Mali Population'. Available at: www.worldometers.info/world-population/mali-population/. (Accessed: 1 March 2017).

Worldometers. (2017c) 'Niger Population'. Available at: www.worldometers.info/world-population/niger-population/. (Accessed: 1 March 2017).

Youssef, P. (2016.) 'Violence around Lake Chad is leading nowhere – just deepening divides', *The Guardian*, 30 June. Available at: www.theguardian.com/global-development/2016/jun/30/violence-around-lake-chad-is-leading-nowhere-just-deepening-divides. (Accessed: 23 February 2017).

6 Findings and concluding remarks

Findings

Conflicts over water resources appear to be a major source of direct international conflict. The most common environmental elements around which conflicts can erupt are population, water flow, diversion, salinisation, floods, climatic variations, and pollution (Schwartz and Singh, 1999, p. 34). Moreover, the institutional capacity to adapt to climatic variability and growing demand is inherent to the emergence of conflict.

The key to adapting to climatic variability, population growth, and the construction of large dams lies in the capacity of institutions. Low levels of institutional development have a more difficult time trying to adapt (Earle, 2005, p. 57). It is also important to note that these key issues influence each other. According to Muller *et al.* (2015, p. 103),

> There is a persistent refrain in the shared waters discourse about the danger that conflicts will arise because of competition over scarce water.

However, a substantial number of international river basins are confronted by inherent climatic variability, a naturally uneven river flow, and a growing demand for water resources because of population growth. These types of rivers characterise Africa. In addition, the region has a history of political instability, a major factor affecting its institutional capacity to deal with transboundary issues. As a result, Africa's transboundary rivers and associated ecosystems could become sources of conflict.

Several findings can be drawn. First, the book underlined that the scarcity of vital resources, and the absence of institutional capacity, can contribute to environmental conflicts. The book was successful, through Homer-Dixon's environmental scarcity theory, in showing that water can be subject to environmental scarcity, thus leading to environmental conflicts. The scarcity of water and its contribution to the intensity of the competition over resources has therefore presented a new landscape of global conflict that requires close and careful analysis. It is in this argument that it becomes evident that water, just as any other economic resource, has the potential to cause tensions and conflict

DOI: 10.4324/9781003366379-6

among users. Three cases [e.g., the Orange Senqu River Basin (OSRB), Nile River Basin (NRB), and the Niger River] were essential in supporting these arguments.

Second, the challenges of population growth, water scarcity, environmental degradation, and insecurities as a result of the uneven distribution of water resources exposed the likelihood of conflict in these respective cases. In addition to the challenges of environmental scarcity, is a lack of adequate cooperative measures to consolidate water sustainability, management, and use. The volatility and proximity of transboundary basins due to the high dependence of all riparian countries on water resources is also a salient feature of nascent environmental conflicts.

The book found that environmental scarcity can contribute to the outbreak of conflict in the absence of institutions. Poor or weak institutions are catalysts to the emergence of environmental conflicts in regions where environmental scarcity has reached boiling point. This is because societies and states become highly vulnerable to the effects of environmental scarcity, with no means, proper assistance, or both to adapt. As a result, the inability to adapt to the impact of environmental scarcity will enable issues, such as migration and resource capture as a means of survival, which subsequently causes degradation, depletion, and an uneven distribution of resources, thus leading to conflict.

Third, the book further found that institutional support structures are important factors for the sustainability, management and use of the Orange Senqu, Nile, and Niger River basins. Institutions are important mechanisms in mediating state behaviour over a shared resource, thus avoiding conflict. This is a notion that has been supported by various researchers (Keohane and Nye, 1977; Krasner, 1983; Godana, 1985; Young, 1986; Keohane, 1989; Kukk and Deese, 1996; Homer-Dixon, 1999a and b; Hasenclever *et al.*, 2000; Okoth-Owiro, 2004; Raadgever *et al.*, 2008; Lindemann, 2009; Neumayer, 2001).

Moreover, the book found that the role of external actors, particularly China, is significant for the sustainability, management, and use of transboundary basins. As a result, these external actors can be integrated into existing theories (Homer-Dixon's environmental scarcity and regime theories) as factors influencing the nature and politics of transboundary basins.

Transboundary and water-related issues, such as the provision of safe drinking water, the provision of water for food production and safety, access to adequate sanitation, cooperation in transboundary water basins (since water resources are transboundary), developing sustainable hydropower to enhance energy security while simultaneously protecting the environment, meeting growing water demands, preventing land degradation and water pollution, and managing water under global climate change can be adequately addressed by institutions.

Regime capacity in the OSRB and RNB is promising since there is a level of cooperation among riparian states and an acceptance of regime authority. This makes the development potential of these basins real and the avoidance of a nascent conflict in these basins possible. The positive outlook is a result of participation by riparian states in institutions created to govern transboundary basins. In contrast to the NBI, ORASECOM and the NBA have achieved a sense of full participation and membership of riparian countries, thus giving these regimes a legitimate, constituent, and legal position to be operational over these basins. This means that the environmental challenges faced by these basins have the potential of being addressed in the near future.

Although the ORASECOM and NBA continue to face markedly high levels of environmental scarcity challenges threatening their capacity and stability, their regime capacity will remain resilient against the effects. The book draws support for this idea from the institutional resiliency argument. According to Solomon (2014, p. 89), the establishment of cooperative water regimes is key to ensuring the resilience of regimes. This is also made possible, as argued earlier, by full riparian membership and participation.

Concluding remarks

Africa is faced with various water-related issues that pose serious challenges to the peace and security discourse on the continent. These challenges include:

- The provision of safe drinking water.
- The provision of water for food security.
- Access to adequate sanitation.
- Cooperation in transboundary water basins (since water resources are transboundary).
- Developing sustainable hydropower to enhance energy security while simultaneously protecting the environment.
- Meeting growing water demands.
- Preventing land degradation and water pollution.
- Managing water under global climate change.
- Enhancing regime capacity to address water-related issues.

These challenges, as mentioned earlier in the book, pose risk factors for a nascent environmental conflict between riparian states in the face of weak regimes.

This book conceptualised environmental scarcity to highlight the potential of conflict in transboundary river basins. Environmental scarcity is triggered by a combination of population growth and excessive strain on dwindling renewable resources, exacerbated by unequal access to that resource (Schwartz *et al.*, 2000, p. 79).

Furthermore, the book conceptualised regimes, particularly on transboundary basins, as treaties on international rivers that hold essential norms and encourage rule-based cooperation to politically resolve problems and conflict in the field of international river basin management (Lindemann, 2009, pp. 700–701).

The OSRB is certainly of major economic importance to South Africa and Lesotho, as it contributes to each country's GDP by 26% and 100%, respectively (Naidoo, 2017). However, environmental issues, notably land degradation in Lesotho, puts the country's water resources and, subsequently, the OSRB under threat (*Lesotho Times*, 2017). This is a very important issue because Lesotho is where the river rises. There seems to be a lack of coordination and domestication of land and water policies in Lesotho.

Lesotho places the entire region at risk by not adequately managing and sustaining environmental resources or resolving issues as far as the harmonisation and coordination of riparian states' environmental policies are concerned. OSRB riparian states will face difficulty in coordinating their water policies as Lesotho itself has difficulties coordinating its policy planning related to its land and water resources. These difficulties minimise the potential for basin-wide policy coordination and integration. According to Water Affairs Minister Kimetso Mathaba,

> coordination and integration planning in Lesotho could be achieved by instigating the integrated catchment management approach.

However, Lesotho adopted the Water and Sanitation Policy of 2007, the Water Act of 2008 and the Long-Term Water and Sanitation Strategy in 2016, as indicated in Chapter 3. These all present a positive outlook for Lesotho's water sector (*Lesotho Times*, 2017). Moreover, as a result of the Climate Resilient Water Resources Investment Strategy and Multipurpose Project Preparation, highlighted in Chapter 3, Lesotho and its population will benefit from this project since its population lives within the basin (Naidoo, 2017).

ORASECOM's agreement with the African Water Facility and the New Partnership for Africa's Development Infrastructure Project Preparation Facility (NEPAD-IPPF) will assist it to develop an optimised water resources investment plan and select a priority transboundary project that will be prepared at the feasibility level. Having integrated climate resilience into the process makes it possible for ORASECOM to holistically address environmental scarcity challenges. The agreement further allows the basin to enhance water resource management, more investments, and more multipurpose projects (Naidoo, 2017). This forms part of the water regimes that are essential for basin management, as argued by the regime theory (Lindemann, 2009, pp. 700–701).

This book argues that in contrast to the Nile and Niger river basins, the water resource developments in the OSRB will improve because multifaceted initiatives are taken, thus improving livelihoods and engendering sustainable socioeconomic growth in the region. It is therefore evident that the Orange Senqu region is highly developed compared with the Nile and Niger, which lack strong regional river basin institutions and are faced with markedly high levels of environmental scarcity.

Although there are institutions, agreements, or both that may perpetuate the uneven use, distribution, or both OSRB, they do not structurally prohibit the use of the basin by other riparian states as is the case with the NRB. However, as far as the emergence of environmental conflicts over water resources is concerned, they are not likely in the immediate future. Theoretically, although environmental scarcity in the form of demand, supply, and structural-induced scarcities is widespread as posited by Homer-Dixon's environmental scarcity theory; regime theory indicated regime formation in these basins that showed signs of effectiveness, to mediate the effects of environmental scarcity.

The case of the NRB validated the probability of environmental conflicts in the immediate future in that, given the reduced outputs due to population growth, degradation and depletion of the Nile, and the uneven distribution of the Nile, competition over the Nile increases the potential for an interriparian conflict in the NRB. The theoretical underpinnings of the book found that environmental scarcity and ineffective regime capacity can lead to environmental conflicts; hence, Homer-Dixon's environmental scarcity approach and regime theory are key to explaining the current state of the Nile's scarcity, institutional capacity, and conflict.

Supporting the idea that, as a result of environmental scarcity, a nascent environmental conflict in the NRB is probable is Homer-Dixon (1999, p. 139), who argued that for a possible conflict to arise in a shared river basin, the following aspects must hold in a narrow set of circumstances:

- First, the downstream countries' national security or survival must depend heavily on the constant flow of the river in its direction.
- Second, the attitude of upstream countries must be threatening to the flow of the river.
- Third, the relations between upstream and downstream countries must be unpleasant.
- Fourth, the downstream countries must be militarily stronger than the upstream countries.
- Finally, fearing that upstream countries may limit water flow or use water as leverage, the downstream countries may use intimidation or direct force.

The NRB, among other shared river basins, provides the best example in which all these conditions hold. First, Egypt is heavily dependent on the water of the Nile; so much so that it has regarding the free flow of the river as a matter of national security (Hadebe, 2013). Second, Egypt perceives the construction of dams in Ethiopia, for instance, as a hindrance or threat to the flow of the Nile's water (Al-Labbad, 2013).

Third, tensions exist between downstream Sudan and Egypt, and upstream Burundi, Eritrea, Ethiopia, Kenya, Rwanda, Tanzania, Uganda, the DRC, and the recently founded South Sudan, and battle lines have been drawn (Hadebe, 2013). According to Starrs (1991, p. 8), Egypt and Ethiopia share a long history of verbal attacks against one another over the Nile.

Fourth, Egypt holds absolute control over the running of the NRB and is militarily stronger than the other states, allowing it to dictate the terms and conditions of the basin's use. Egypt controls the region's most powerful military. Finally, fearing that its upstream neighbours will reduce its water supply by constructing dams, Egypt has warned that it is ready to use force to protect its access to the Nile (Simpkins, 2010).

These events in the NRB reveal the central role that the Nile's water plays in high-stakes geopolitical developments (Stoa, 2014, pp. 1353–1354). The increase in China's involvement in the NRB further complicates the situation. China is a key external actor and a willing alternative financial donor for upstream countries and has been vital in the power shift and in shaping the landscape of transboundary water governance and development along the Nile (Heinrich Böll Foundation, 2012).

China has become a major exporter of hydropower projects to the Nile region (Erdal, 2013). Chinese companies and banks are currently funding and undertaking numerous hydraulic projects in upstream countries, such as Burundi, Ethiopia, Kenya, Tanzania, the DRC, Uganda, and Sudan, making it the fastest-growing player in Nile developments (Bosshard, 2009, p. 47).

The reason for China's involvement in the NRB is to assist upstream countries to develop the Nile, thus changing the Nile's hydropolitical landscape, which has been dominated by Egypt for most of the last century. Egypt controlled and constructed the hydropolitical regime to favour its interests and maintain its hegemonic position in the basin. This was mainly made possible by its material, bargaining and ideational power (Cascão, 2009, p. 248). Hanke (2013, pp. 59–68) argued that Egypt's powerful and influential status, for example, its economic development, military might, political stability, and access to external political and financial support vis-a-vis upstream countries, enabled it to influence and construct the Nile discourse.

One of Egypt's most common and successful means of influencing the Nile discourse was its power to prevent financial institutions, regionally and internationally, from funding upstream countries' hydropower infrastructures (Waterbury, 2002, p. 80). Donors, such as the World Bank, and European Investment Bank, and regional banks, such as the African Development Bank have been openly reluctant to fund projects in upstream countries due to Egyptian influence, stating,

> all downstream countries, Egypt in particular, have to concede to it for any projects to be financed,
>
> (Green Prophet, 2010; Jacob, 2012, p. 138; Mayton, 2012)

China has become an alternative source of funding to upstream countries' hydropower projects, which has subsequently changed the hydropolitical status quo of the Nile. However, the presence of China in the Nile dam industry has presented positive and negative effects on the Nile and its surrounding environment. The positive effects include upstream countries being able to use the Nile and develop hydropower projects. The negative effects include the creation of more environmental and social uncertainties through the financial and technical support of these unilateral projects. Homer-Dixon's environmental scarcity theory helped outline these uncertainties of human and environmental security and showed how these projects aggravated environmental scarcity.

In a 2005 BBC interview, the former UN Secretary-General, Boutros Boutros-Ghali, argued,

> the real problem is that we need an additional quantity of water and we will not have an additional quantity of water unless we find an agreement with the upstream countries which also need water and have not used the Nile water until now,
>
> (Thompson, 2005)

It is been made clear from the book's findings that a major problem surrounding the NRB is the lack of a binding agreement that ensures that all 11 riparian countries share common benefits from the Nile River (Turton, 2000b, p. 2).

According to Othieno and Zondi (2006, pp. 2–3):

> There has to be a resolution to the question of balance between who has the historic right to the Nile's water, and issues of sovereign water rights especially with regards to where the river ought to be controlled, whether that be upstream or downstream. There also has to be a resolution on the impact of improper environmental and social assessments in the construction of hydropower projects that ultimately lead to massive displacements while degrading and depleting the Nile.

Considering the long-anticipated or predicted conflicts in the NRB, binding agreements should be put in place. Agreements, institutional structures, or both would be essential in this regard as they would forge a new international water agreement for the NRB. This requires a more serious interventionist role by the African Union (AU) and the NBI because, whether conflict emerges or not largely depends on regime capacity in the region. The AU may help analyse state policies to identify common interests.

The AU Commission has been involved in federating river and lake basin authorities under the aegis of the African Network of Basin Organisations. It has developed policy and institutional framework guidelines about cooperation for the sustainable management of transboundary water basins. The Guidelines for the Establishment of the Cooperative Framework Agreement (CFA) for the Integrated Management of Transboundary Basins have been developed and disseminated to member states of the AU.

The AU's future involvement in water diplomacy efforts in the region may centre on promoting regional cooperation and providing a forum for discussion. The AU can help facilitate exchange and cooperation among scientific and technological experts and provide a forum for transparent discussions between the parties.

However, the AU has played a limited role in past water diplomacy efforts between Egypt and Ethiopia. In addition, the AU's failure to change the status quo in the basin and achieve an equitable use of the Nile further maintained

Egypt's veto power. The NBI played an equally limited role as far as the NRB downstream-upstream conflict is concerned.

First, the NBI arguably did little to address the fundamental schism over the options regarding rights and obligations that differ between upstream and downstream riparian countries. Second, relatively little appears to have been done about the problematic water use by certain Nile riparian countries. These included unsustainable practices, such as the development of new settlements in the desert, the farming of relatively water-intensive crops, and storing water in reservoirs with relatively high evaporation rates.

Third, the process for negotiating the NBI and the CFA has arguably not been sufficiently inclusive. For the most part, only government officials from the Nile riparian countries and representatives from the World Bank, UNDP or CIDA, or both have attended the negotiations. Very few additional stakeholders have been allowed to participate in the negotiations and opportunities for public involvement have not been substantial, all of which made the CFA ineffective.

Furthermore, the fact that a comprehensive cooperative framework over the management and use of the Nile is ineffective means that the occurrence of disputes and environmental stress on the Nile is inevitable. The AU and the NBI need to rectify the current Nile environmental issues by forging a cooperative framework to enable a common position on the management and equitable exploitation of the Nile. Their role should take into consideration the Nile region's already complex political environment.

Other than financial assistance, the international community (e.g., the US, UN, and the AU) can play an important role in the negotiations, advocating for collaborative efforts for effective and sustainable water management by all parties in the NRB.

In the RNB, where water scarcity and flow variability are always cause for concern, the only option for the sustainability of the water resources, optimal use, and good member relationships is to pursue the path of coordinated, cooperative water resources development. It is only the development of dams that perpetuates the uneven use, distribution, or both in the RNB. Similar to the OSRB, and in contrast to the NRB, agreements and institutions established in the RNB weren't created to prohibit the use of the Niger River by other riparian countries. Moreover, they do not structurally deny other riparian countries access to the Niger River.

In the first three phases of the institutional and legal framework of the NBA, a challenge confronting the Niger basin was the establishment of an effective transboundary organisation over 20 years. The political will and financial responsibility of member states were central to these challenges, and both these key issues can be summed up as a lack of accountability.

Since 1998, the NBA has experienced a strong commitment from the political leadership of member countries; the payment of membership dues has improved owing to the new formula adopted, and the Secretariat has been strengthened by recruiting top managers and technical experts and through more regular meetings of ministers and heads of states. These positive signals have brought an increase in donor support.

To conclude these three case studies, it is important to note that while environmental changes may play a role in setting the stage for conflict, the causal relationship is highly complex and the role of regimes, institutions, or both are crucial in the emergence of conflict.

In accordance with regime theory, for regimes to be successful and viable in these three case studies, the three additional key requirements as outlined by Andersen *et al.* (2005, p. 59) (e.g., legitimacy, relevance, and constituency) are crucial. Other important factors, such as capacity and financial viability, are often a result of a lack of these three prerequisites.

First, the legitimacy of an institution is defined by its legal and juridical basis and is also a function of how the institution is perceived by those it is intended to serve, its credibility among stakeholders, the level of competency of its staff, and the transparency of its governance. Second, the relevance of the institution is similarly significant. Finally, it is the shareholders of the institution, the constituency, who will be the ultimate judge of whether the river basin organisation has achieved legitimacy and relevance. Unless the constituency sees the NBA as directly relevant to domestic development priorities, it is unlikely to place real value on the institution (Andersen *et al.*, 2005, p. 59).

Andersen *et al.* (2005, p. 59) further argued that a practical consequence will be that the annual budgets of the institution are left unpaid and will become increasingly irrelevant to national priorities, initiating a downward spiral that will be exacerbated as the institution, to maintain itself, will seek resources from a variety of sources (including donors), involving activities that no longer respect the institution's mandate.

It is important to draw an analysis of the interaction and effects of demand, supply, and structural-induced scarcity in the OSRB, NRB, and RNB, respectively to determine environmental scarcity outcomes. One major reason to assess this interaction is because Homer-Dixon (1994, pp. 8–11) earlier asserted that the environmental scarcity theory stipulated that these types of scarcity (e.g., demand, supply, and structural-induced scarcity) are not mutually exclusive. They often occur simultaneously and interact, ultimately leading to conflict. Furthermore, Schwartz *et al.* (2000, p. 80) indicated,

> uneven distribution never acts on its own, its impact is always a function of its interaction with resource demand and supply.

Based on Homer-Dixon's (1991, p. 91) argument, the OSRB, NRB, and the RNB are expected to face a number of effects as a result of the occurrence, interaction, or both of demand, supply, and structural-induced scarcity. These include decreased agricultural potential, regional economic decline, population displacement, and the disruption of legitimised and authoritative institutions and social relations.

From the arguments raised in this book, the continuation of demand, supply, and structural-induced scarcity in the OSRB, NRB, and the RNB will affect agricultural potential, which will subsequently reduce regional economic activity as a result of worsening deforestation, water scarcity, and environmental degradation and depletion. Following a decline in economic activity, agricultural potential, and low per capita access to water in specific areas will eventually lead to massive population displacements. This phenomenon of displacement will be further exacerbated by the construction of large dams and inter and intra basin water transfers (IBTs).

Moreover, these aspects will all contribute to the failure, weakening, collapse, or both of authoritative institutions that are meant to help societies adapt to scarcity and minimise the potential of violence. This will lead to conflict between the riparian countries of the OSRB, NRB, and RNB. It was therefore essential that an analysis of regimes governing the OSRB, NRB, and RNB was provided to determine their effectiveness and capacity to help states and societies adapt to the effects of environmental scarcity and thus avoid a nascent environmental conflict.

With or without these challenges, Africa must use and share its transboundary resources collectively, equitably, and profitably for the benefit of its present and future social, economic, and political needs. It is only through such cooperative efforts that cooperation can be enhanced and interstate tension and conflict avoided.

Bibliography

Al-Labbad, M. (2013) 'Egypt, Ethiopia headed for war over water', *Al-Monitor*, 24 March. Available at: www.al-monitor.com/pulse/political/2013/03/egypt-ethiopia-water-war.html. (Accessed: 14 May 2013).

Andersen, I. *et al.* (2005) *The RNB: A vision for sustainable management*. Washington, DC: The International Bank for Reconstruction and Development / the World Bank.

Bosshard, P. (2009) 'China dams the world', *World Policy Journal*, 26(4).

Cascão, A.E. (2009) 'Changing power relations in the Nile River basin: Unilateralism vs. cooperation?' *Water Alternatives*, 2(2).

Earle, A. (2005) 'Hydropolitics in Southern Africa: What is the prognosis for peaceful development of shared watercourses?', FWU, Vol.3, Topics of integrated Watershed Management – Proceedings, Summer School.

Erdal, Z. (2013) 'Strengthening new bonds: China's investments' role in boosting Ethiopia industrialisation', *Consultancy Africa*, 4 February. Available at: www.consultancyafrica.com/index.php?option=com_content&view=article&id=1209:strengthening-new-bonds-chinese-investments-role-in-boosting-ethiopian-industrialisation-&catid=87:african-finance-a-economy&Itemid=294. (Accessed: 16 March 2015).

Godana, B.A. (1985) *Africa's Shared Water Resources*. Boulder, Colorado: Lynne Rienner Publishers.

Green Prophet. (2010) 'Upstream African countries sign Nile Water deal against Egypt's interests', 18 May. Available at: www.greenprophet.com/2010/05/upstream-african-countries-sign-nile-water-deal-against-egypts-interest/. (Accessed: 30 October 2013).

Hadebe, S.P. (2013) 'Egypt's total monopoly of the Nile River affects development potential of the lower stream riparian states', *Jaywalking Political Discussions*, 28 April. Available at: http://shadebe.wordpress.com/2013/04/18/egypts-total-monopoly-of-the-nile-river-affects-development-potential-of-the-lower-stream-riparian-states/. (Accessed: 14 May 2013).

Hanke, N. (2013) '*East Africa's growing power: Challenging Egypt's hydropolitical position on the Nile?*' Unpublished thesis. Stellenbosch University.

Hasenclever, A., Mayer, P., and Rittberger, V. (2000) 'Integrating theories of international regimes', *Review of International Studies*, 26.

Heinrich Böll Foundation. (2012) *Sudan after separation: New approaches to a new region*. Available at: www.ke.boell.org/sites/default/files/sudan_after_separation_web.pdf. (Accessed: 8 February 2017).

Homer-Dixon, T.F. (1991) 'On the threshold: Environmental changes as causes of acute conflict', *International Security*, 16(2).

Homer-Dixon, T.F. (1999b) 'Threshold of turmoil: Environmental scarcities and violent conflict', in Deudney, D.H. and Matthew, R.A. (eds.) *Contested Grounds: Security and Conflict in the New Environmental Politics*. Albany: State University of New York Press, pp. 61-90.

Homer-Dixon, T.F. (1994) 'Environmental scarcities and violent conflict: Evidence from cases', *International Security*, 19(1).

Homer-Dixon, T.F. (1999a) *Environment, scarcity, and violence*. New Jersey: Princeton University Press.

Jacobs, I.M. (2012) *The Politics of Water in Africa: Norms, Environmental Regions and Transboundary*. Cooperation in the Orange-Senqu and the Nile Rivers. New York: Bloomsbury Academic.

Keohane, R.O., and Nye, J.S. (1977) *Power and interdependence*. Boston: Little, Brown.

Keohane, R.O. (1989) *International institutions and state power, essays in international relations theory*. Boulder, CO: Westview Press.

Krasner, S.D. (1983) *International regimes*. Ithaca, NY: Cornell University Press.

Kukk, C., and Deese, D. (1996) 'At the water's edge: Regional conflict and cooperation over fresh water', *UCLA Journal of International Law and Foreign Affairs*, 21(1).

Lesotho Times. (2017) 'Water sources under threat', 25 February. Available at: http://lestimes.com/water-sources-under-threat/. (Accessed: 13 March 2017).

Lindemann, S. (2009) 'Success and failure of international river basin management: The case of southern Africa', in Brauch, H.G. *et al.* (eds.) *Facing global environmental change: Environmental, Human, Energy, Food, Health and Water Security concepts*', Hexagon Series on Human and Environmental Security and Peace, 4, pp. 699-709. Berlin: Springer.

Mayton, J. (2012) 'Water for all: The Nile Basin conundrum', *DANTEmag*, 8 December. Available at: www.dantemag.com/2012/12/water-for-all-the-nile-basin-conundrum/. (Accessed: 30 October 2013).

Muller, M., Chikozho, C., and Hollingworth, B. (2015) 'Water and regional integration: The role of water as a driver of regional economic integration in Southern Africa', Water Research Commission, Report No 2252/1/14.

Naidoo, R. (2017) 'Orange-Senqu River Basin gets R48.6m for water security', Infrastructurene.ws. 20 February. Available at: www.infrastructurene.ws/2017/02/20/orange-senqu-river-basin-gets-r48-6m-for-water-security/#. (Accessed: 14 March 2017).

Neumayer, E. (2001) 'How regime theory and economic theory of international environmental cooperation can learn from each other', *Global Environmental Politics*, 1(1).

Okoth-Owiro, A. (2004) 'The Nile Treaty: State succession and International Treaty Commitments: A case study of the Nile water Treaties'. Nairobi: Konrad Adenauer Foundation.

Othieno, T., and Zondi, S. (2006) 'Co-operative management of Africa's water resources: a basis for resolving conflicts on the Nile River Basin,' *Institute for Global Dialogue*, 5(59).

Raadgever, G.T. *et al.* (2008) 'Assessing management regimes in transboundary river basins: Do they support adaptive management?' *Ecology and Science*, 13(1).

Schwartz, D., and Singh, A. (1999) 'Environmental conditions, resources, and conflicts: An introductory overview and data collection', Nairobi: UNEP.

Schwartz, D.M., Deligiannis, T., and Homer-Dixon, T.F. (2000) 'The environment and violent conflict: A response to Gleditsch's critique and some suggestions for future research'. *Environmental change & security, project report*', 6. Summer.

Simpkins, G. (2010) 'Africa faces water war threats', *Africa Rising 21st Century*, 26 February. Available at: http://africarising2010.blogspot.co.za/2010/02/africa-faces-water-war-threats.html. (Accessed: 11 March 2016).

Solomon, H. (2014) 'Potential for cooperation rather than conflict in the face of water degradation: The cases of the Nile River and Okavango River Basins', *Southern Journal for Contemporary History*, 39(1).

Starr, J. R. (1991) 'Water wars'. *Foreign Policy*, 82(Spring).

Stoa, R.B. (2014) 'The United Nations Watercourses Convention on the dawn of entry into force', *Vanderbilt Journal of Transnational Law*, 47.

Thompson, M. (2005) 'Ex-UN chief warns of water wars', BBC News, 2 February. Available at: http://news.bbc.co.uk/2/hi/africa/4227869.stm. (Accessed: 21 March 2013).

Turton, A.R. (2000b) *A cryptic hydropolitical history of the Nile Basin for the students of hydropolitics*', Study Guide for Pretoria University, Pretoria: African Water Issues Research Unit (AWIRU).

Waterbury, J. (2002) *The Nile basin national determinants of collective action*. New Haven: Yale University Press.

Young, O.R. (1986) 'International regimes: Toward a new theory of institutions', *World Politics*, 39(1).

Index

Note: Page numbers in bold refers to tables. Endnotes are indicated by the page number followed by "n"

For Product Safety Concerns and Information please contact our EU
representative GPSR@taylorandfrancis.com
Taylor & Francis Verlag GmbH, Kaufingerstraße 24, 80331 München, Germany

www.ingramcontent.com/pod-product-compliance
Lightning Source LLC
Chambersburg PA
CBHW071353220926
51613CB00009B/188
* 9 7 8 1 0 3 2 4 3 2 4 3 4 *